110 Operational Amplifier Projects for the Home Constructor

R. M. MARSTON

Newnes Technical Books

THE BUTTERWORTH GROUP

UNITED KINGDOM Butterworth & Co (Publishers) Ltd
London: 88 Kingsway, WC2B 6AB

AUSTRALIA Butterworths Pty Ltd
Sydney: 586 Pacific Highway, Chatswood, NSW 2067
Also at Melbourne, Brisbane, Adelaide and Perth

CANADA Butterworth & Co (Canada) Ltd
Toronto: 2265 Midland Avenue, Scarborough, Ontario, M1P 4

NEW ZEALAND Butterworths of New Zealand Ltd
Wellington: 26–28 Waring Taylor Street, 1

SOUTH AFRICA Butterworth & Co (South Africa) (Pty) Ltd
Durban: 152–154 Gale Street

USA Butterworth (Publishers) Inc
Boston: 19 Cummings Park, Woburn, Mass. 01801

First published 1975

Second impression published by Newnes Technical Books
a Butterworth imprint 1976

ISBN 0 408 00153 4

Printed in England by Billing & Sons Limited,
Guildford, London and Worcester

CONTENTS

PREFACE

Of the many new semiconductor devices introduced to the electronics world in the past decade, one of the most important and versatile is a device known as the operational amplifier, or 'op-amp'. The modern op-amp is a high-gain d.c. differential amplifier, having a high input and low output impedance, and is readily available in integrated circuit form. They have a multitude of applications in the home and in industry, and can readily be used as the basis of a host of a.c. and d.c. amplifiers, instrumentation circuits, oscillators, tone generators, and sensing circuits, etc.

This book is intended to be of equal interest to the electronics amateur, student, and engineer. With this aim in mind, the volume starts off by outlining the essential characteristics of the op-amp, and then goes on to show 110 useful projects in which the devices can be used. All of these projects have been designed, built, and fully evaluated by the author, and range from simple amplifiers to sophisticated instrumentation circuits. The operating principle of each project is explained in concise but comprehensive terms, and brief constructional notes are given where necessary.

The volume is designed to be of interest to both English and American readers, and all projects have been designed around the internationally available type 709 and 741 operational amplifiers. All other semiconductors used in the circuits are equally popular and readily available international types. As an aid to construction, the outlines of all semiconductors used in the projects are given in the volume appendix. Unless otherwise stated, all resistors used in the projects are standard half-watt types.

CHAPTER 1

BASIC PRINCIPLES AND APPLICATIONS

Op-amps were originally designed to perform the mathematical operations of addition, subtraction, integration, etc., in analogue computers. The devices have many other uses, however, and can readily be used as the basis of a host of a.c. and d.c. amplifiers, instrumentation circuits, oscillators, tone generators, and sensing circuits, etc. In this present volume we show 110 different projects that can be built around these versatile devices.

Basic characteristics and circuits

Most operational amplifiers are of the differential-input type, and are represented by the symbol shown in *Figure 1.1a*. *Figure 1.1b* shows the basic supply connections that are used with an op-amp. Note that the device is operated via a dual power supply with a common ground, thus enabling the op-amp output to swing either positive or negative with respect to ground.

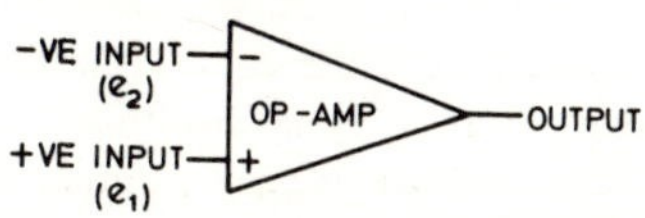

Figure 1.1a. Basic op-amp symbol.

The op-amp has two input terminals, and uses direct coupling between input and output. Typically, the device gives a basic low-frequency voltage gain of about 100 000 between input and output, has an input impedance of about 1 MΩ at each input terminal, and has an output impedance of a few hundred ohms.

One input terminal of the device is denoted negative, and gives an inverted output, and the other is denoted positive, and gives a non-inverted output. If a positive input voltage is applied to the negative

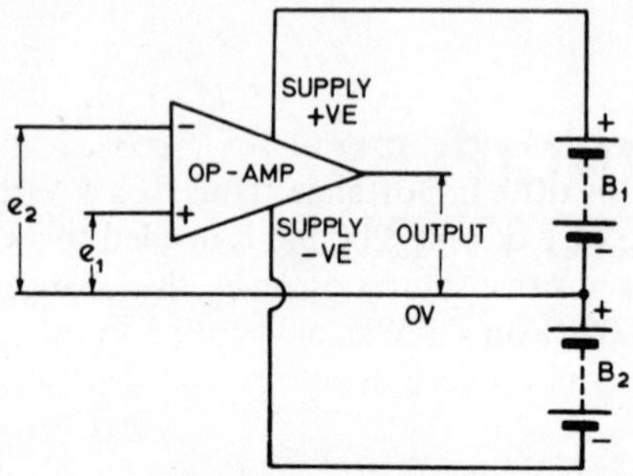

Figure 1.1b. Basic supply connections of an op-amp.

terminal while the other input is grounded the output is inverted, and swings negative. Alternatively, if a positive input is applied to the positive terminal while the other terminal is grounded the output is non-inverted, and swings positive. If identical signals are simultaneously applied to both inputs the output will ideally be zero, since the two signals are cancelled out by the differential action of the amplifier. Note that the output of the circuit is proportional to the *differential* signal between the two inputs, and is given by:

$$e_{out} = A_0(e_1 - e_2)$$

where A_0 = the open-loop voltage gain of the op-amp (typically 100 000).

e_1 = signal voltage at the positive input.

e_2 = signal voltage at the negative input.

Figure 1.2a shows a very simple application of the op-amp. This particular circuit is known as a differential voltage comparator, and has a fixed reference voltage applied to the negative input terminal, and a

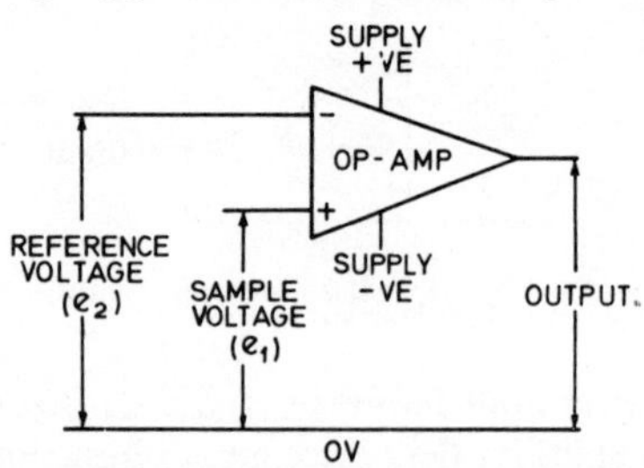

Figure 1.2a. Simple differential voltage comparator circuit.

variable test or sample voltage applied to the positive terminal. When the sample voltage is greater than that of the reference by more than a few hundred microvolts the output is driven to saturation in the positive direction, and when the sample is greater than a few hundred microvolts less than the reference voltage, the output is driven to saturation in the negative direction.

Figure 1.2b shows the voltage transfer characteristics of the above circuit. Note that it is the magnitude of the differential input voltage that dictates the magnitude of the output voltage, and that the absolute values of input voltage are of little importance. Thus, if a 1 V reference is used and a differential voltage of only 200 μV is needed to switch the output from a negative to a positive saturation level, this change can be caused by a shift of only 0.02 % on a 1 V signal applied to the sample input. The circuit thus functions as a precision voltage comparator or balance detector.

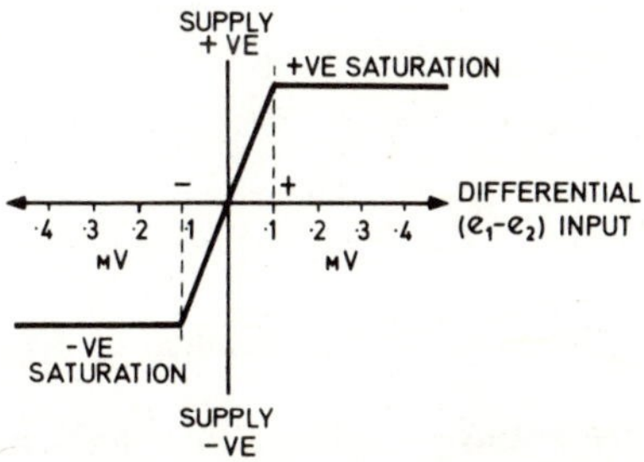

Figure 1.2b. Transfer characteristics of the differential voltage comparator circuit of Figure 1.2a.

The op-amp can be made to function as a low-level inverting d.c. amplifier by simply grounding the positive terminal and feeding the input signal to the negative terminal, as shown in *Figure 1.3a*. The op-amp is

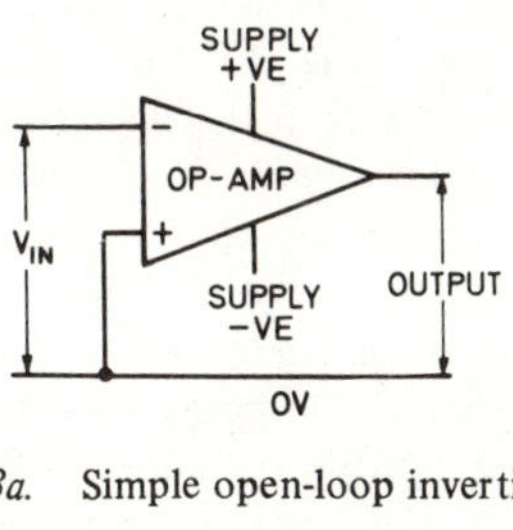

Figure 1.3a. Simple open-loop inverting d.c. amplifier.

used 'open-loop' (i.e., without feedback) in this configuration, and thus gives a voltage gain of about 100 000 and has an input impedance of about 1 MΩ. The disadvantage of this circuit is that its parameters are dictated by the actual op-amp, and are subject to considerable variation between individual devices.

A far more useful way of employing the op-amp is to use it in the closed-loop mode i.e., with negative feedback. *Figure 1.3b* shows the method of applying negative feedback to make a fixed-gain inverting d.c. amplifier. Here, the parameters of the circuit are controlled by feedback

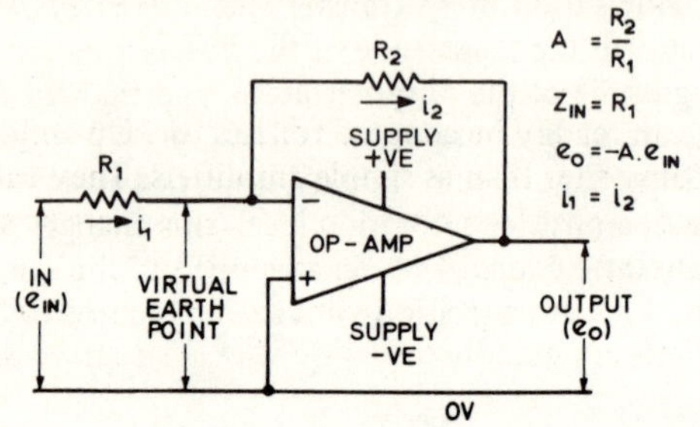

Figure 1.3b. Basic closed-loop inverting d.c. amplifier.

resistors R_1 and R_2. The gain, A, of the circuit is dictated by the ratios of R_1 and R_2, and equals R_2/R_1. The gain is virtually independent of the op-amp characteristics, provided that the open-loop gain (A_0) is large relative to the closed-loop gain (A). The input impedance of the circuit is equal to R_1, and again is virtually independent of the op-amp characteristics.

It should be noted at this point that although R_1 and R_2 control the the gain of the complete circuit, they have no effect on the parameters of the actual op-amp, and the full open-loop gain of the op-amp is still available between its negative input terminal and the output. Similarly, the negative terminal continues to have a very high input impedance, and negligible signal current flows into the negative terminal. Consequently, virtually all of the R_1 signal current also flows in R_2, and signal currents i_1 and i_2 can be regarded as being equal, as indicated in the diagram.

Since the signal voltage appearing at the output terminal end of R_2 is A times greater than that appearing at the negative terminal end, the current flowing in R_2 is A times greater than that caused by the negative terminal signal only. Consequently, R_2 has an apparent value of R_2/A when looked at from its negative end, and the $R_1 - R_2$ junction thus appears as a low-impedance *virtual earth* point.

It can be seen from the above description that the *Figure 1.3b* circuit is very versatile. Its gain and input impedance can be very precisely controlled by suitable choice of R_1 and R_2, and are unaffected by variations in the op-amp characteristics. A similar thing is true of the non-inverting d.c. amplifier circuit shown in *Figure 1.4a.* In this case the voltage gain is equal to $(R_1 + R_2)/R_2$, and the input impedance is approximately equal to $(A_0/A)Z_{in0}$, where Z_{in0} is the open-loop input impedance of the op-amp. A great advantage of this circuit is that it has a very high input impedance.

The op-amp can be made to function as a precision voltage follower by connecting it as a unity-gain non-inverting d.c.amplifier, as shown in *Figure 1.4b.* In this case the input and output voltages of the circuit are identical, but the input impedance of the circuit is very high and is approximately equal to $A_0 \times Z_{in0}$.

The basic op-amp circuits of *Figure 1.2a* to *1.4b* are shown as d.c. amplifiers, but can readily be adapted for a.c. use. Op-amps also have many applications other than as simple amplifiers. They can easily be made to function as precision phase splitters, as adders or subtractors, as active filters or selective amplifiers, as precision half-wave or full-wave rectifiers, and as oscillators or multivibrators, etc. A whole range of useful applications are described in following chapters of this volume.

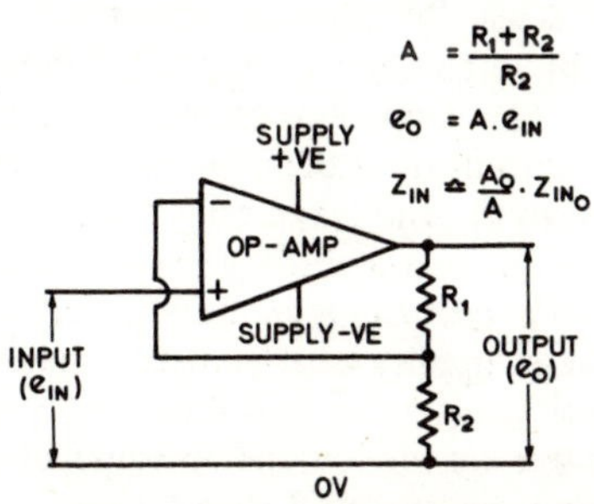

Figure 1.4a. Basic non-inverting d.c. amplifier.

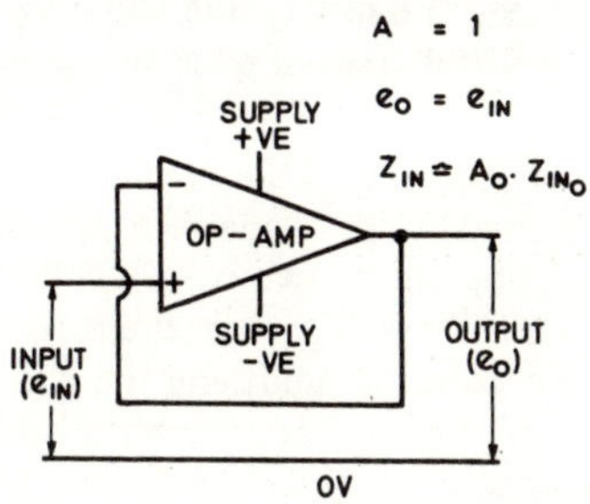

Figure 1.4b. Basic unity-gain d.c. voltage follower.

Op-amp parameters

An ideal operational amplifier would have an infinite input impedance and zero output impedance, would have infinite gain and infinite bandwidth, and would give perfect tracking between input and output.

Practical op-amps fall far short of the ideal, and have finite gain, bandwidth, width, etc., and give tracking errors between the input and output signals. Consequently, various performance parameters are detailed on op-amp data sheets, and indicate the measure of 'goodness' of the particular device type in question. The most important of these parameters are detailed below.

Open-loop voltage gain, A_o This is a measure of voltage gain occurring directly between the input and output terminals of the op-amp, and may be expressed in direct terms or in terms of dB. Typical gain figures of modern op-amps are 100 000, or 100 dB.

Input impedance, Z_{in} This is a measure of the impedance looking directly into the input terminals of the op-amp, and is usually expressed in terms of resistance only. Values of 1 MΩ are typical of modern op-amps.

Output impedance, Z_o This is a measure of the output impedance of the basic op-amp, and is usually expressed in terms of resistance only. Values of one or two hundred ohms are typical of modern op-amps.

Input bias current, I_b Most op-amps use bipolar transistor input stages, and draw a small bias current from the input terminals. The magnitude of this current is denoted by I_b, and is typically only a fraction of a microamp.

Supply voltage range, V_s Op-amps are usually operated from two sets of supply rails, and these supplies must be within maximum and minimum limits. If the supply voltages are too high the op-amp may be damaged, and if the supply voltages are too low the op-amp will not function correctly. Typical supply limits are ± 3 V to ± 15 V.

Input voltage range, $V_{i(max)}$ The input voltage to the op-amp must never be allowed to exceed the supply line voltages, or the op-amp may be damaged. $V_{i\,(max)}$ is usually specified as being one or two volts less than V_s.

Output voltage range, $V_{o(max)}$ If the op-amp is over driven its output will saturate and be limited by the available supply voltages, so $V_{o\,(max)}$ is usually specified as being one or two volts less than V_s.

Differential input offset voltage, V_{io} In the ideal op-amp perfect tracking would exist between the input and the output terminals of the device, and the output would register zero when both inputs were grounded. Actual op-amps are not perfect devices, however, and in practice slight imbalances exist within their input circuitry and effectively cause a small offset or bias potential to be applied to the input terminals of the op-amp. Typically, this *differential input offset voltage* has a value of only a few millivolts, but when this voltage is amplified by the gain of the circuit in which the op-amp is used it may be sufficient to drive the op-amp output to saturation. Because of this, most op-amps have some facility for externally nulling out the offset voltage.

Common mode rejection ratio, c.m.r.r. The ideal op-amp produces an output that is proportional to the difference between the two signals applied to its input terminals, and produces zero output when identical signals are applied to both inputs simultaneously, i.e., in common mode. In practical op-amps, common mode signals do not entirely cancel out, and produce a small signal at the op-amps output terminal. The ability of the op-amp to reject common mode signals is usually expressed in terms of common mode rejection ratio, which is the ratio of the op-amps gain with differential signals to the op-amps gain with common mode signals. C.M.R.R. values of 90 dB are typical of modern op-amps.

Transition frequency, f_T An op-amp typically gives a low-frequency voltage gain of about 100 dB, and in the interest of stability its open-loop frequency response is tailored so that the gain falls off as the frequency rises, and falls to unity at a transition frequency denoted f_T. Usually, the response falls off at a rate of 6 dB per octave or 20 dB per decade. *Figure 1.5* shows the typical response curve of an op-amp with an f_T of 1 MHz and a low frequency gain of 100 dB.

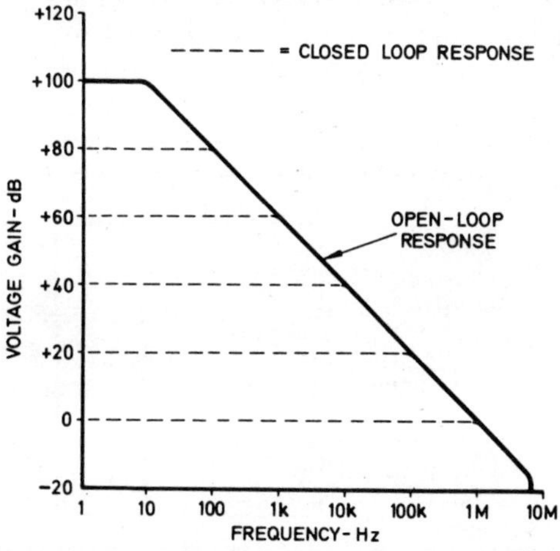

Figure 1.5. Typical op-amp frequency response curve.

Note that, when the op-amp is used in a closed-loop amplifier circuit, the bandwidth of the circuit depends on the closed-loop gain. If the amplifier is used to give a gain of 60 dB its bandwidth will be only 1 kHz, and if it is used to give a gain of only 20 dB its bandwidth will extend to 100 kHz. The f_T figure can thus be used to represent a gain-bandwidth product.

Slew rate, S As well as being subject to normal bandwidth limitations, op-amps are also subject to a phenomenon known as slew rate limiting, which has the effect of limiting the maximum rate of change of voltage at the output of the device. Slew rate is normally specified in terms of volts per microsecond, and values in the range 1 V/μs to 10 V/μs are common with the most popular types of op-amp. One effect of slew rate limiting is to make a greater bandwidth available to small output signals than is available to large output signals. Another effect is to convert sine wave input signals into triangle wave output signals when the op-amp is operated beyond its slew rate.

Power supplies for op-amps

Op-amps require the use of two power supply sources for satisfactory operation. One of these supplies must be positive relative to the common input signal point, and the other must be negative. In most applications these supplies are obtained by using two independent supply sources connected at a common point, as shown in the circuit of *Figure 1.1b*. Normally, these supplies are of the balanced types, in which the supply voltages are equal in magnitude but opposite in polarity. It should be noted, however, that the use of balanced supplies is not mandatory, and unbalanced supplies can be used in cases where the maximum possible symetrical peak-to-peak output signal is not required from the op-amp.

It is not essential to use two independent supplies to provide the two power sources for the op-amp, since two power sources can be obtained from a suitably adapted single power supply unit. *Figure 1.6a* shows one

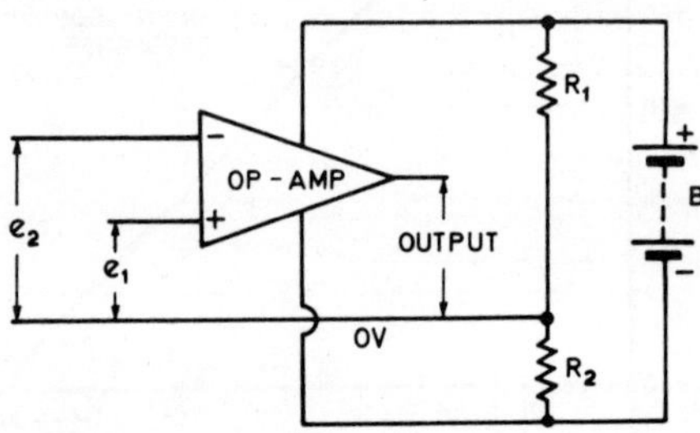

Figure 1.6a. **Potential divider method of powering an op-amp from a single supply source in d.c. applications.**

method of obtaining the supplies from a single power unit. Here, potential divider $R_1 - R_2$ is wired across the single supply, and the $R_1 - R_2$ junction is used as the common signal point, thus making a positive supply rail available at the top of R_1 and a negative supply rail available at the bottom of R_2. In d.c. applications the values of R_1 and R_2 must be chosen so that the quiescent current flowing through them is much greater than the peak output current that is to be taken from the op-amp output, since these resistors are effectively in series with the op-amp output.

In cases where the op-amp is to supply a high peak output current the above requirement may result in the need for unacceptably high quiescent currents in R_1 and R_2. One way round this problem is to replace R_1 and R_2 with a zener diode potential divider, as shown in *Figure 1.6b*. The zener diodes present a low dynamic impedance in series with the op-amp output, so in this case their quiescent currents need be only slightly greater than the peak output current of the op-amp, and can be adjusted via R_1.

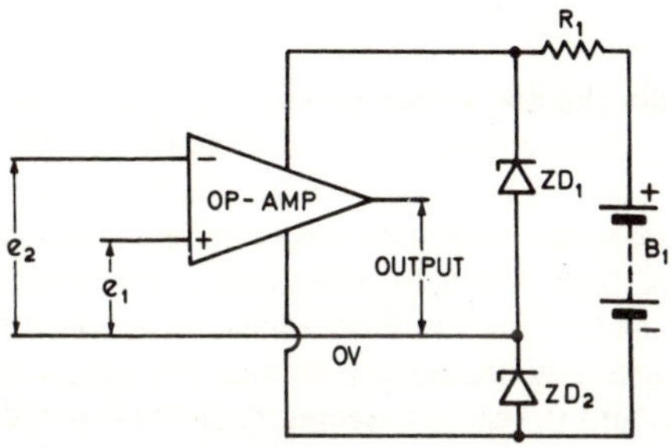

Figure 1.6b. Zener potential divider method of powering an op-amp from a single supply source in d.c. applications.

The two single-supply circuits that we have looked at so far are designed to power d.c. amplifiers, and need to pass fairly high quiescent currents because both the signal and the supply currents are d.c. and flow through common resistive elements. In the case of a.c. circuits alternative supply networks can be used, and quiescent currents can be much lower.

Figure 1.7 shows one method of powering an a.c. op-amp circuit from a single power unit. Here, potential divider $R_1 - R_2$ is again wired across

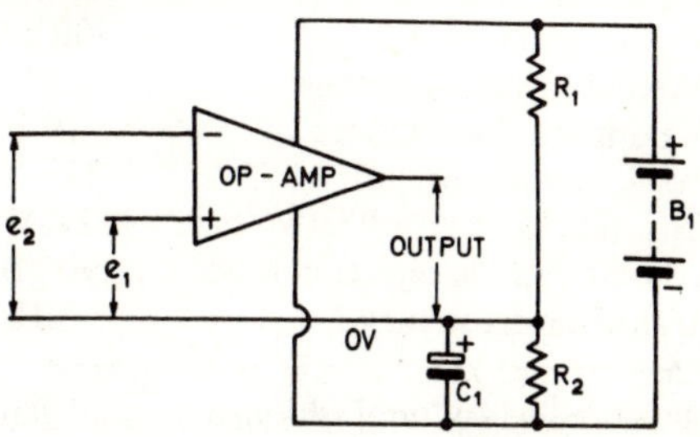

Figure 1.7. Method of powering an op-amp from a single supply source in a.c. applications.

the single supply unit, and the $R_1 - R_2$ junction is used to act as the common signal point, but in this case R_2 is shunted by large-value capacitor C_1. Consequently, a very low a.c. impedance exists between the common signal line and the negative supply rail (via the low impedance of C_1), and between the common signal line and the positive supply rail (via the low internal impedance of supply unit B_1 in series with C_1), and the a.c. current-driving ability of the op-amp is thus not influenced by the values or quiescent currents of R_1 and R_2. In fact, the only current-related requirement of R_1 and R_2 is that their quiescent currents be large relative to the input bias current (I_b) parameter of the op-amp, and in most cases quiescent currents of only a few microamps can be used.

Practical op-amps: The 709 and the 741

Many types of operational amplifier are commercially available. Some are specifically designed to have exceptional high-frequency parameters, some are designed to give exceptionally high input impedances or to exhibit exceptional thermal stability, and some are designed simply for general purpose use. Two of the best known general purpose types are the 709 and the 741, and the main parameters of these two devices are listed in *Table 1.1.* The 709 and 741 op-amp types are available from a number of manufacturers, under a variety of codings and in a variety of packagings.

Table 1.1 Typical characteristics of the 709 and 741 operational amplifiers.

	Parameter	*709*	*741*
A_o	Open-loop voltage gain	93 dB	100 dB
Z_{in}	Input impedance	250 kΩ	1 MΩ
Z_o	Output impedance	150 Ω	150 Ω
I_b	Input bias current	300 nA	200 nA
$V_{s(max)}$	Maximum supply voltage	±18 V	±18 V
$V_{i(max)}$	Maximum input voltage	±10 V	±13 V
$V_{o(max)}$	Maximum output voltage	±14 V	±14 V
V_{io}	Differential input offset voltage	2 mV	2 mV
c.m.r.r.	Common mode rejection ratio	90 dB	90 dB
F_T	Transition frequency	5 MHz	1 MHz

The 709 op-amp is a slightly old-fashioned 'second generation' operational amplifier. It has a number of design weaknesses, but is still widely used. The device is subject to a phenomenon known as *input latch up*, in which the input circuitry may switch into a locked state if special precautions are not taken when connecting the input signals to the input terminals, and the op-amp can easily be destroyed by short circuits inadvertently placed across the output terminals. In addition, the device is prone to bursting into unwanted oscillations when used in the linear mode, and makes use of external frequency compensation components for stability control. A major advantage of the 709 op-amp is that it has a higher slew rate and better bandwidth than the 741 op-amp. In the present volume the 709 is used in only a few circuits, and in these is used purely in a switching capacity, so that the high slew rate is utilised without incurring the disadvantages that accrue when the device is used in the linear mode.

The 741 op-amp is a greatly improved 'third generation' version of the 709 op-amp. It is immune to input latch up, has a short circuit proof output, and has built-in frequency compensation and is not prone to instability when used in the linear mode. The frequency response characteristics of the device are identical to those shown in *Figure 1.5*, and the unity gain bandwidth is typically 1 MHz. The device can be fitted with external offset nulling by wiring a 10 kΩ pot between its two null terminals, and taking the pot slider to the negative supply rail, as shown in *Figure 1.8*.

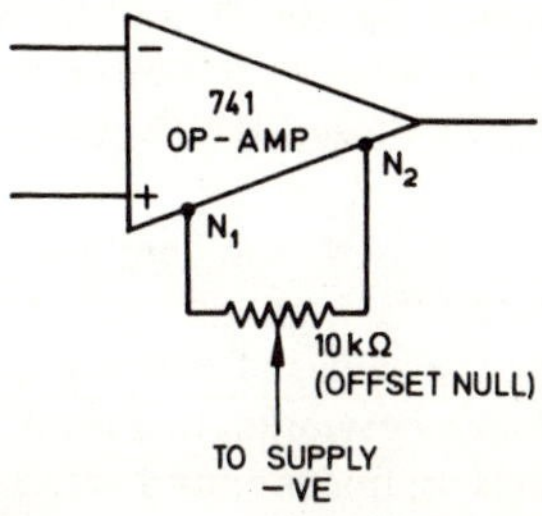

Figure 1.8. Method of applying offset nulling to the type 741 operational amplifier.

All one hundred and ten of the circuits described in the following chapters of this volume are designed around the type 741 op-amp, and the pin connections shown in each of the respective circuit diagrams apply to the 8-pin dual-in-line version of the device only. If alternatively packaged 741 op-amps are used in these circuits, the pin connections may have to be changed. A variety of 741 pin connection arrangements are shown in the appendix to this volume.

CHAPTER 2

25 A.C. AND D.C. AMPLIFIER PROJECTS

The high open-loop voltage gains and direct couplings of operational amplifiers enable the devices to be used in a wide variety of d.c. and a.c. amplifier applications. Because of the differential input facility of the op-amp, such amplifiers can be designed to be of either the inverting, the non-inverting, or the differential types.

When op-amps are used as closed-loop amplifiers the amplifier characteristics can, because of the high inherent gain of the op-amp, be dictated almost entirely by the values of external feedback components. By suitably selecting feedback networks, therefore, op-amps can readily be persuaded to act as precision linear amplifiers, as non-linear amplifiers, as frequency-selective amplifiers, or as constant-volume amplifiers, etc.

Twenty-five useful d.c. and a.c. amplifier projects of various types are shown in the present chapter. All of these circuits are designed around the popular type 741 integrated-circuit op-amp, and the pin connections shown in the following diagrams apply to the 8-pin dual-in-line version of this device only.

Inverting amplifier projects

An op-amp can be made to function as an inverting amplifier by grounding the positive input terminal and feeding the input signal to the negative terminal. If the amplifier is used in the open-loop mode the circuit will give a low-frequency voltage gain of about 100 000, and an input signal of a millivolt or so will be sufficient to drive the output to saturation. If the op-amp is used in the closed-loop mode, on the other hand, the circuit gain will be dictated by the values of the external feedback components, and almost any required values of voltage gain and input impedance can be obtained.

Figure 2.1a shows the connections for making an inverting d.c. amplifier with a voltage gain of 100, or 40 dB. Here, feedback resistor R_2 is wired between the op-amp output and the negative input terminal, and the input signal is applied to the negative input via R_1. The positive terminal is grounded via R_3.

There are two important facts to remember when looking at this circuit. First, the actual op-amp has a very high input impedance (typically 1 MΩ), so very little signal current flows into the negative input of the op-amp. The second point to remember is that the op-amp has a typical open-loop gain of 100 000 times. With these points in mind, consider the effect of R_2 on the circuit.

R_2 is wired as a negative feedback resistor between the output and the negative input terminal of the op-amp. Consequently, if an input of 100 μV is connected to the negative side of R_2, 10 volts will appear at the output and thus across R_2. The negative feedback thus effectively reduces the value of R_2 to R_2/A_{vo} where A_{vo} is the open-loop voltage gain of the op-amp. This modified resistance is effectively in parallel with the open loop input resistance of the op-amp, so the negative input appears as a 'virtual ground' low-impedance point.

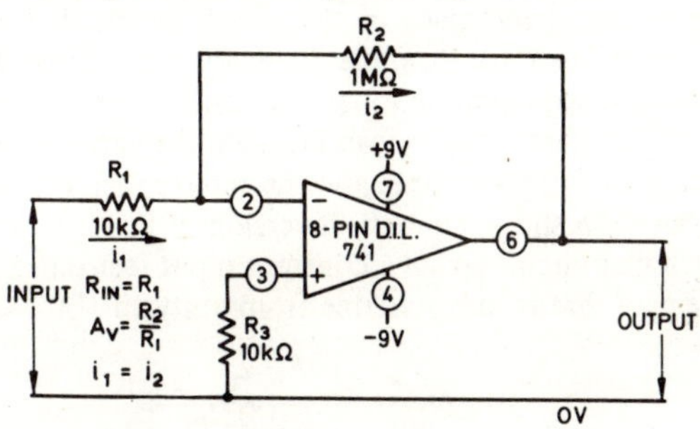

Figure 2.1a. x 100 inverting d.c. amplifier.

Although R_2 changes the input resistance of the amplifier, it has no effect on the voltage gain of the actual op-amp. The gain of the circuit (as opposed to the gain of the op-amp) is, however, changed by wiring R_1 in series between the circuits input terminal and the input of the op-amp. In this case R_1 and the 'virtual ground' resistance act effectively as a potential divider which causes only a fixed fraction of the input signal to be applied to the input of the op-amp, so reducing the gain of the overall circuit. The actual voltage gain, A_v, of the circuit works out at

$$A_v = \frac{R_2}{R_1 + \dfrac{R_2}{A_{vo}}}$$

In practice this formula simplifies to $A_v = R_2/R_1$ since A_{vo} is very large. The voltage gain of the *Figure 2.1a* circuit works out at $10^6/10^4 = 100$. Note that the voltage gain is dictated purely by the values of R_1 and R_2, and is virtually independent of variations in the op-amp characteristics.

There are three further points to note about this circuit. First, since the negative input terminal of the op-amp acts as a virtual ground, the input resistance of the circuit is equal to R_1. Hence, the basic circuit can be designed to give any required values of input resistance and voltage gain by choosing suitable values for R_1 and R_2.

The second point to note is that, since negligible current flows into the high-impedance negative terminal of the actual op-amp, any signal current that flows in R_1 must also flow in R_2, and signal currents i_1 and i_2 are thus equal.

Finally, note that the value of the R_3 resistor that is wired between the positive input and ground is chosen to give optimum thermal-drift performance of the op-amp, and should have a value equal to the parallel resistance of R_1 and R_2.

The *Figure 2.1a* circuit is designed to give a fixed voltage gain. The circuit can be modified and made to give a variable gain in a number of alternative ways. R_1 can, for example, be made a variable resistor, in which case both the gain and the input resistance can be varied simultaneously. Alternatively, R_2 can be made the variable resistor, in which case the gain will be variable and the input resistance will be constant. *Figure 2.1b* shows a practical version of this last-mentioned type, this particular circuit giving a constant input resistance of 10 kΩ amd a voltage gain that is fully variable from unity to 100.

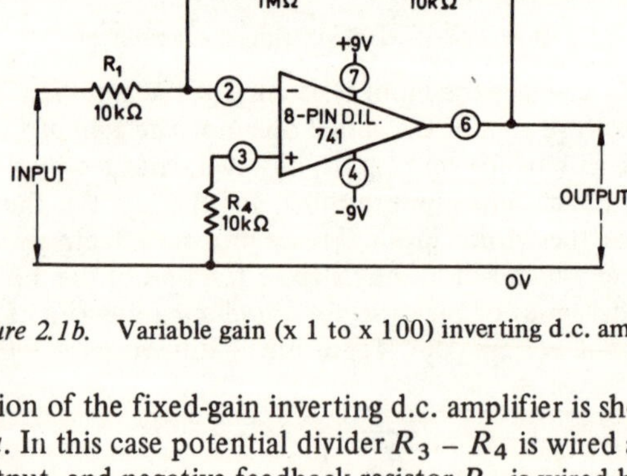

Figure 2.1b. Variable gain (x 1 to x 100) inverting d.c. amplifier.

A variation of the fixed-gain inverting d.c. amplifier is shown in *Figure 2.2a*. In this case potential divider $R_3 - R_4$ is wired across the op-amp output, and negative feedback resistor R_2 is wired between the $R_3 - R_4$ junction and the negative input terminal. This configuration enables both R_1 and R_2 to be given high values while still giving high voltage gain. The voltage gain is given by

$$A_v = \frac{R_2}{R_1} \times \frac{R_3 + R_4}{R_4}$$

The *Figure 2.2a* circuit has an input resistance of 1 MΩ, and a voltage gain of 100.

The *Figure 2.2a* circuit can be made to give a variable gain in any one of a number of ways. The gain can be made variable by changing the value of any one of the four resistors, or by replacing $R_1 - R_2$ or

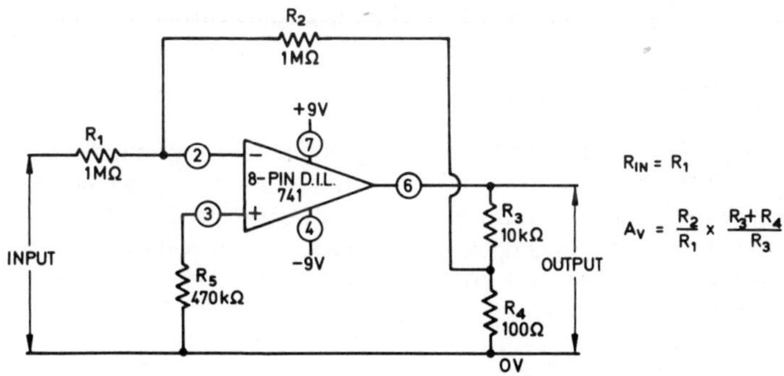

Figure 2.2a. High-impedance x 100 inverting d.c. amplifier.

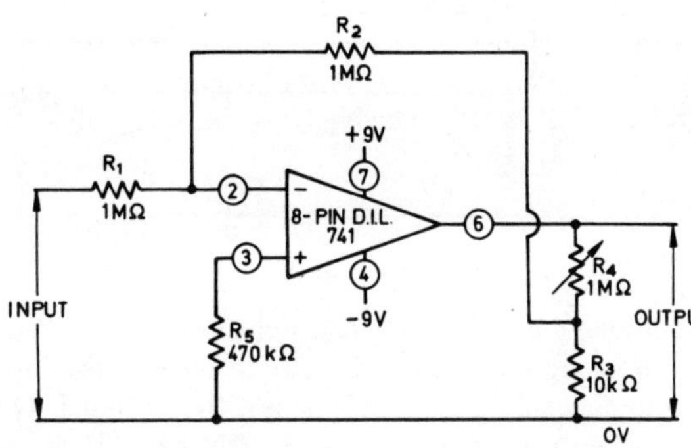

Figure 2.2b. High-impedance, variable gain (x 1 to x 100) inverting d.c. amplifier.

$R_3 - R_4$ by a variable potential divider. *Figure 2.2b* shows how the gain can be varied via R_4, while retaining a constant input resistance of 1MΩ to the amplifier.

The inverting circuits shown so far are used as d.c. amplifiers. They can readily be modified for a.c. use by simply wiring blocking capacitors in series with their inputs and outputs, as shown in the fixed gain inverting a.c. amplifier of *Figure 2.3*.

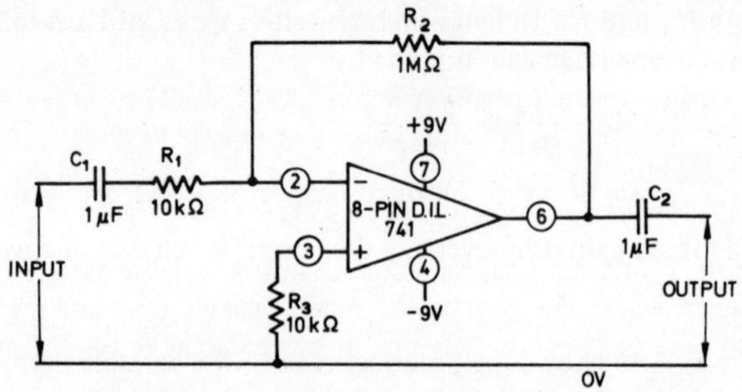

Figure 2.3. x 100 inverting a.c. amplifier.

Non-inverting amplifier projects

An op-amp can be made to function as a non-inverting amplifier by feeding the input signal to the positive terminal and applying negative feedback to the negative terminal via a resistive potential divider that is connected across the op-amp output. *Figure 2.4a* shows the connections for making a fixed gain (x 100) d.c. amplifier.

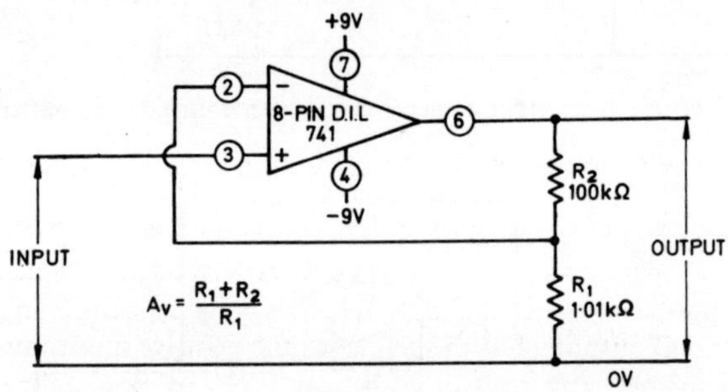

Figure 2.4a. Non-inverting x 100 d.c. amplifier.

Here, potential divider $R_1 - R_2$ is wired across the op-amp output, and the $R_1 - R_2$ junction is taken directly to the negative input of the op-amp; the input signal is applied to the positive terminal. The output signal is in phase with the input, and the voltage gain, A_v, is related to the values of R_1 and R_2 by the formula

$$A_v = \frac{R_1 + R_2}{R_1}$$

Hence, if R_2 is given a value of zero the gain falls to unity, and if R_1 is given a value of zero the gain rises towards infinity (but in practice is limited to the open-loop gain of the op-amp). The gain of the *Figure 2.4a* circuit works out at 100.

A major advantage of the non-inverting d.c. amplifier is that it gives a very high input impedance to the positive terminal. In theory the input resistance is equal to the open-loop input resistance (typically 1 MΩ) multiplied by the open-loop voltage gain (typically 100 000) divided by the actual circuit voltage gain. In practice input resistance values of hundreds of megohms can readily be obtained.

The basic fixed-gain non-inverting d.c. amplifier circuit of *Figure 2.4a* can be made to give a variable gain by replacing either R_1 or R_2 with a variable resistor, or by replacing R_1 and R_2 with a variable potential divider. *Figure 2.4b* shows the practical circuit of a variable gain d.c. amplifier, in which the gain can be varied from unity to 100 via a variable resistor in the R_2 position.

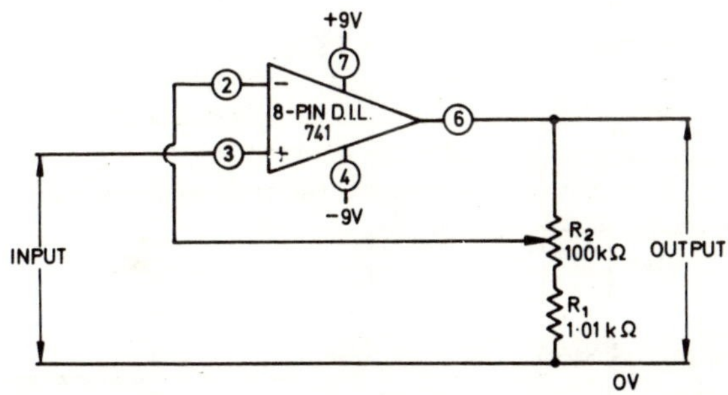

Figure 2.4b. Non-inverting variable gain (x 1 to x 100) d.c. amplifier.

The basic non-inverting d.c. circuits of *Figure 2.4a* and *2.4b* can be modified to operate as a.c. amplifiers in a variety of ways. The most obvious approach here is to simply wire blocking capacitors in series with the inputs and outputs, but in such cases the positive input must be d.c. grounded via a suitable resistor, as shown by R_3 in the fixed-gain non-inverting a.c. amplifier of *Figure 2.5*. If this resistor is not used the op-amp will have no d.c. stability, and its output will rapidly drift into

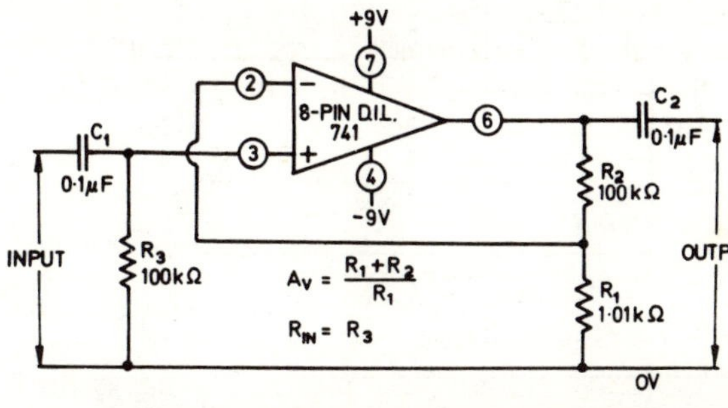

Figure 2.5. Non-inverting x 100 a.c. amplifier.

saturation. Clearly, the input resistance of the *Figure 2.5* circuit is equal to R_3 at operating frequencies, and R_3 must have a relatively low value in the interest of d.c. stability. This circuit thus loses the non-inverting amplifiers basic advantage of high input resistance. The *Figure 2.5* circuit has an input resistance of only 100 kΩ.

A useful development of the *Figure 2.5* circuit is shown in *Figure 2.6*. Here, blocking capacitor C_3 is wired in series with gain-determining potential-divider $R_1 - R_2$, and the $R_2 - C_3$ junction is taken directly to the negative input. The circuit thus has virtually 100 % d.c. negative feedback, gives near-unity d.c. voltage gain, and has excellent d.c. stability. As far as a.c. is concerned, however, C_3 acts as a short circuit, so the circuit gives an a.c. voltage gain of $(R_1 + R_2)/R_1$. Thus, the circuit has a closely controlled a.c. gain, with excellent d.c. stability. The input impedance is equal to R_3, and has a value of 100 kΩ.

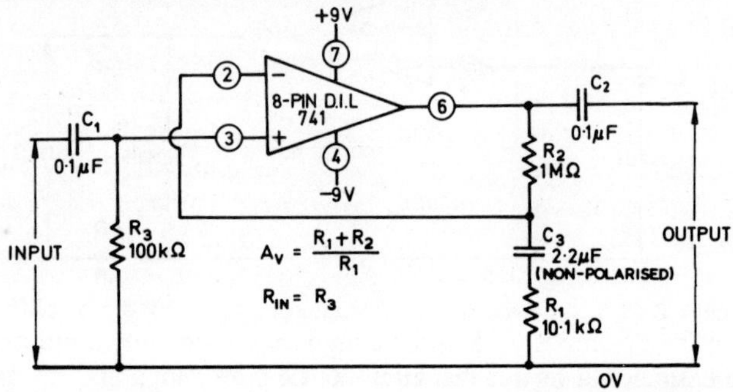

Figure 2.6. Non-inverting x 100 a.c. amplifier with d.c. negative feedback.

The *Figure 2.6* circuit can be further modified, so that it gives a very high input impedance, by using the connections shown in *Figure 2.7*. Here, the low end of input resistor R_3 is taken to the $C_3 - R_1$ junction,

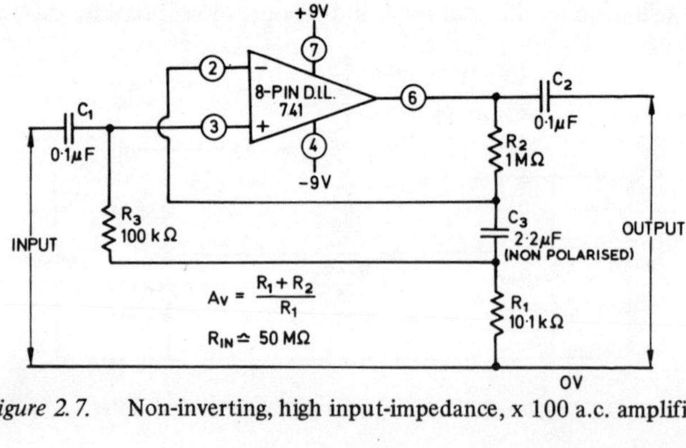

Figure 2.7. Non-inverting, high input-impedance, x 100 a.c. amplifier.

rather than directly to ground. Under a.c. amplifying conditions identical a.c. signals appear at the positive terminal of the op-amp and at the $R_1 - C_3$ junction of the gain-determining potential divider. Identical a.c. signals thus appear at each end of input resistor R_3, so zero signal current flows in this resistor, which consequently appears as a near-infinite resistance to a.c. signals. As a result, the circuit has a very high input resistance (typically of the order of 50 MΩ) as far as a.c. is concerned, but has good d.c. stability due to the fact that a relatively low d.c. resistance path exists between the positive terminal and ground (110 kΩ in this case), and that the circuit has near-unity d.c. gain due to the virtually 100 % d.c. negative feedback that is obtained via R_2.

The use of offset null

The op-amp is a direct-coupled device, and amplifies any d.c. or a.c. signal that appears at its input terminals. Ideally, when the op-amp is used in the open-loop mode, its output should register zero volts when its input terminals are grounded. In practice, however, the output usually goes to saturation under this condition, because internally generated voltages effectively apply a small offset or bias potential to the input circuitry of the op-amp. Typically, this 'differential input offset voltage' has a value of one or two millivolts, and this small d.c. voltage is amplified by the open-loop gain of the op-amp, and drives the output to saturation.

When the op-amp is used in the closed loop mode, the input offset voltage is amplified by a factor equal to the closed loop gain of the circuit. If the op-amp is used as a x 100 d.c. amplifier, and has an input offset potential of 2 mV, an output offset of 200 mV will be obtained when zero volts are applied to the input terminals.

In many applications this offset of the output is undesirable, so most op-amps have some facility for externally nulling or cancelling the effects of the offset voltage. In the case of the 8-pin dual-in-line version of the 741 op-amp, offset nulling is achieved by wiring a 10 kΩ variable potential divider, or pot, between null pins 1 and 5 of the op-amp, and taking the pot slider to the negative supply rail of the circuit. *Figure 2.8* shows the practical connections for applying the offset null facility to a x 100 non-inverting d.c. amplifier. The facility can be applied to any circuit that uses a 741 op-amp, but alternative pin connections may have to be used if types other than the 8-pin d.i.l. version are used.

Voltage follower circuits

An op-amp can be made to function as a precision voltage follower by connecting it as a unity-gain non-inverting amplifier. *Figure 2.9a* shows

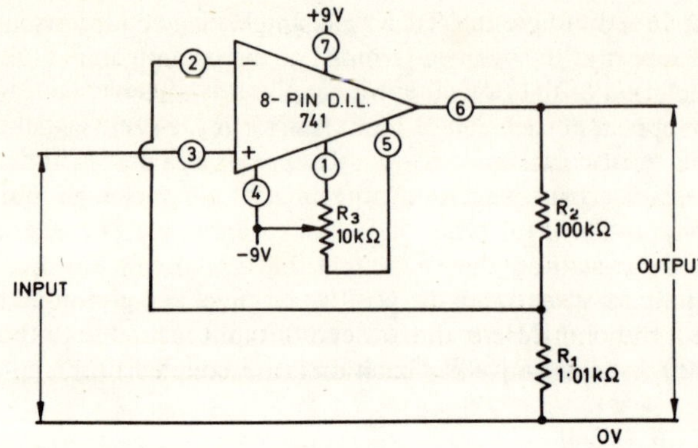

Figure 2.8. Non-inverting x 100 d.c. amplifier with offset null.

the practical connections for making a precision d.c. voltage follower. Here, the input signal is applied directly to the positive terminal of the op-amp, and the negative terminal is connected directly to the output, so the circuit has 100 % d.c. negative feedback and acts as a unity-gain non-inverting d.c. amplifier. The output voltage signal of the circuit is virtually identical to that at the input, so the output is said to 'follow' the input voltage. The great advantage of the circuit is that it has a very high input impedance (as high as hundreds of megohms) and a very low output impedance (as low as a few ohms). The circuit effectively acts as an impedance transformer.

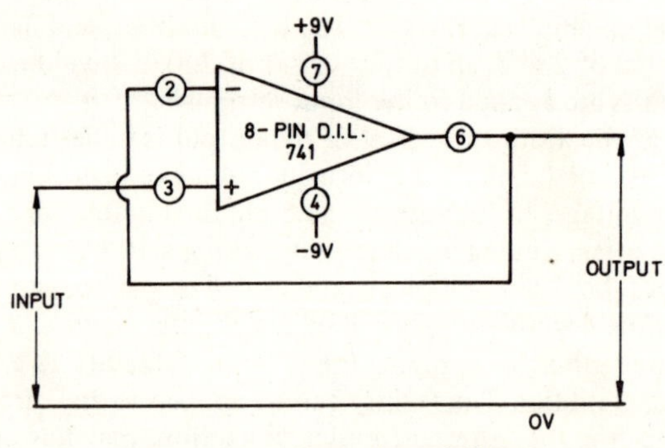

Figure 2.9a. D.C. voltage follower.

In practice the output of the basic *Figure 2.9a* circuit will follow the input to within a couple of millivolts up to magnitudes within a volt or so of the supply line potentials. If required, the circuit can be made to follow to within a few microvolts by adding the offset null facility to the op-amp.

Figure 2.9b shows how the *Figure 2.9a* circuit can be modified so that it acts as an a.c. voltage follower. Here, C_1 is wired in series with the input to block d.c. from the positive terminal, and C_2 is used to block d.c. from the output. R_1 is wired between the positive terminal and ground to provide a discharge path for C_1 and to ensure d.c. stability of the op-amp. Because of the presence of this resistor, the circuit has a resistive input impedance of only 1 MΩ.

Figure 2.9c shows how the a.c. voltage follower circuit can be modified so that it gives a resistive input impedance of hundreds or thousands of megohms. Here, the low end of input resistor R_1 is taken to ground via R_2, and the $R_1 - R_2$ junction is a.c. coupled to the op-amp

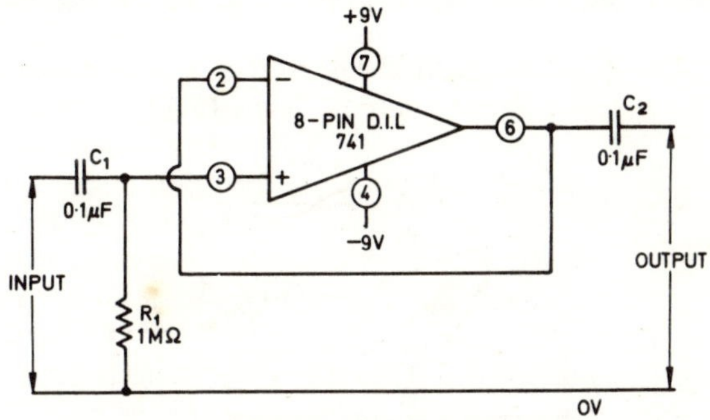

Figure 2.9b. A.C. voltage follower.

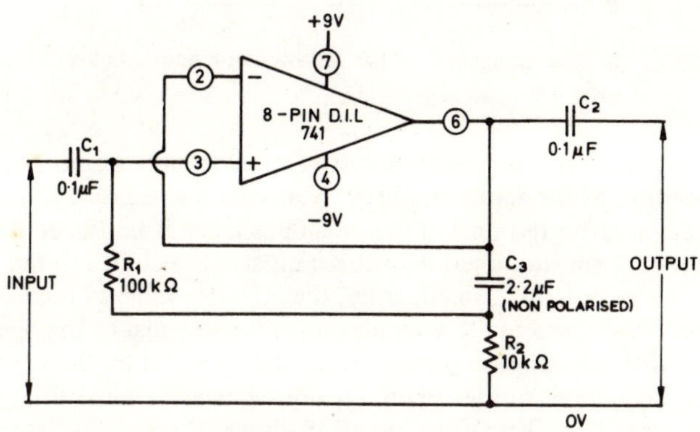

Figure 2.9c. Very high input-impedance a.c. voltage follower.

output via C_3. At a.c. operating frequencies C_3 appears as a virtual short circuit, so the full output signal of the op-amp appears at the $R_1 - R_2$ junction. Since the input and output signals of the circuit are identical, therefore, identical a.c. signals appear at both ends of R_1, and zero signal current flows in this resistor, which thus appears as a near-infinite impedance to a.c. This technique of increasing the apparent value of a resistor is known as *bootstrapping*, and the technique enables the *Figure 2.9c* circuit to exhibit an input impedance of hundreds or thousands of megohms.

The 741 op-amp is capable of providing output currents up to about 5 mA, and this is consequently the current-driving limit of the three voltage follower circuits that we have looked at so far. If required, however, the current-driving capabilities of the circuits can readily be increased by wiring one or more emitter follower buffer stages between the op-amp output terminals and the output of the actual circuit. The precise design of the buffer stage depends on the output requirements of the circuit.

Figure 2.10a shows the practical circuit of an unidirectional d.c. voltage follower with a boosted output. Transistor Q_1 is an npn type,

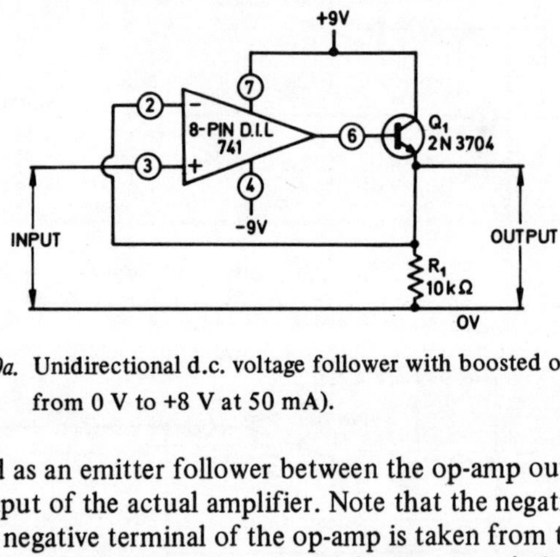

Figure 2.10a. **Unidirectional d.c. voltage follower with boosted output (variable from 0 V to +8 V at 50 mA).**

and is wired as an emitter follower between the op-amp output terminal and the output of the actual amplifier. Note that the negative feedback loop to the negative terminal of the op-amp is taken from the emitter of Q_1, so the base-emitter junction of the transistor is included in the negative feedback loop. Consequently, the effective value of the 600 mV base-emitter volt drop of Q_1 is reduced by a factor equal to the open-loop gain of the op-amp, so Q_1 has no significant effect on the voltage-following capabilities of the circuit. Q_1 does, however, boost the current-driving capability of the circuit to about 50 mA. This figure of 50 mA is dictated by the limited power rating of the 2N3704 transistor. Greater output currents can be obtained by replacing Q_1 with a high-gain power transistor.

Note that this particular circuit is capable of giving a positive output only, since Q_1 is an npn transistor and must be positively biased to operate. The circuit thus acts as an unidirectional voltage follower.

Figure 2.10b shows the practical circuit of a bidirectional d.c. voltage follower with boosted output. This circuit can provide both positive and negative outputs, with currents up to 50 mA mean or about 350 mA peak. Circuit operation is quite simple. $Q_1 - Q_2$ are wired together as a complementary emitter follower so that when the output is positive Q_1 is biased on and provides the output current, and Q_2 is cut off; or when the output is negative, Q_2 is biased on and provides the output current, and Q_1 is cut off: bidirectional outputs are thus available. Note that the base emitter junctions of both transistors are included in the negative feedback loop of the circuit, so these junctions consequently have negligible effect on the voltage following capabilities of the circuit. Slight loss of voltage following capability does, however, occur at near-zero output voltage levels, and may manifest itself in the form of cross-over distortion when the circuit is driven from an a.c. source.

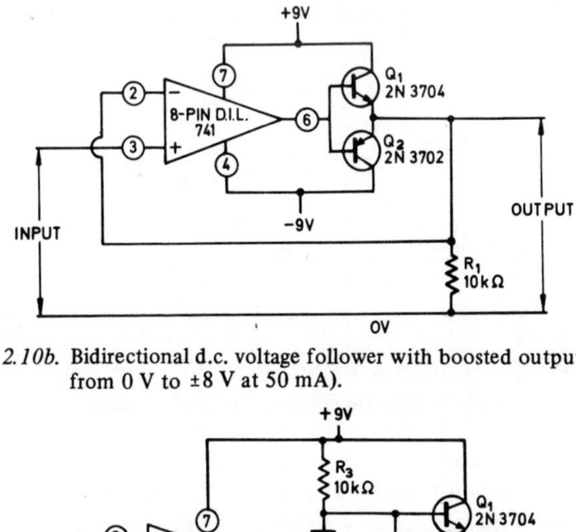

Figure 2.10b. **Bidirectional d.c. voltage follower with boosted output (variable from 0 V to ±8 V at 50 mA).**

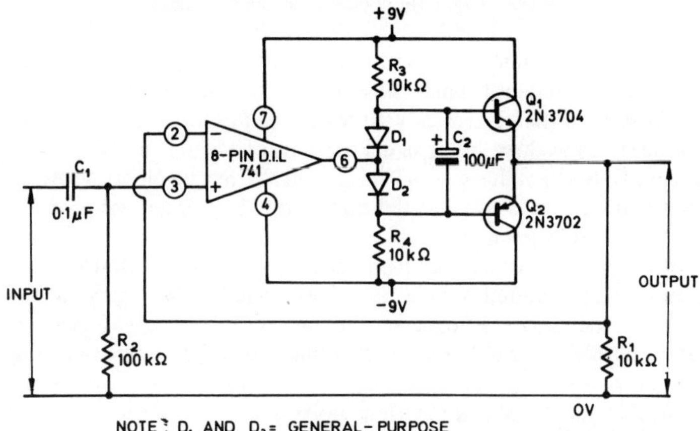

Figure 2.10c. **Improved bidirectional follower with boosted output and a.c. input.**

The improved bidirectional voltage follower circuit of *Figure 2.10c* shows how cross-over distortion can be reduced to negligible proportions by applying a small standing bias to each output transistor via potential divider $R_3 - D_1 - D_2 - R_4$. D_1 and D_2 are general-purpose silicon diodes. The circuit shows the connections to be used with an a.c. input, and C_2 is used to equalise the base drives to Q_1 and Q_2 at normal operating frequencies and so minimise distortion.

Addition circuits

An operational amplifier can be made to carry out the function of addition by connecting it as a multi-input inverting amplifier, as shown in the unity-gain inverting d.c. adder circuit of *Figure 2.11*. Looking at each input network individually, it can be seen that each input resistor combines with negative feedback resistor R_4 to form a unity-gain

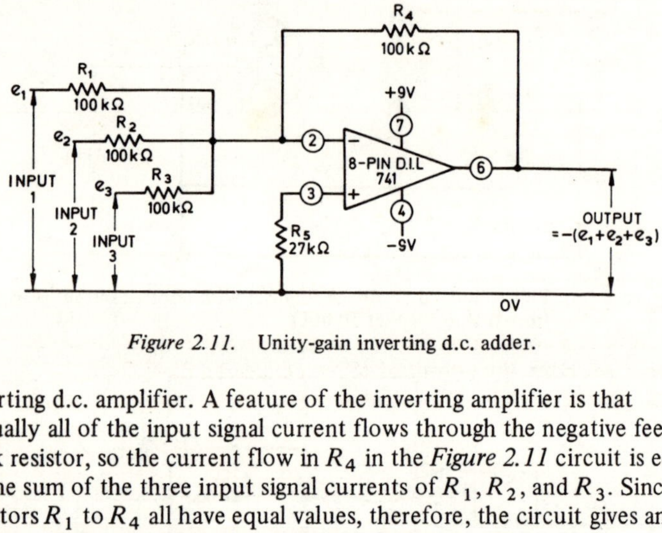

Figure 2.11. Unity-gain inverting d.c. adder.

inverting d.c. amplifier. A feature of the inverting amplifier is that virtually all of the input signal current flows through the negative feedback resistor, so the current flow in R_4 in the *Figure 2.11* circuit is equal to the sum of the three input signal currents of R_1, R_2, and R_3. Since resistors R_1 to R_4 all have equal values, therefore, the circuit gives an output voltage that is equal to the sum of the three input voltages, but is inverted in sign or polarity.

The *Figure 2.11* circuit can be made to give an output that is equal to the sum of any required number of inputs by simply wiring extra input resistors to the circuit. If required, the circuit can be made to give addition with gain by either increasing the value of R_4 or by reducing the values of all of the input resistors: the formula for the voltage gain is $A_v = R_4/R_{in}$, where R_{in} is the input resistor.

The *Figure 2.11* inverting adder circuit can be adapted for a.c. use by wiring blocking capacitors in series with each input resistor and with the output. Adding circuits of this type are widely used as so-called 'mixers' in audio applications, the signal to each input terminal being made adjustable via a variable potential divider.

The *Figure 2.11* circuit gives an output that is inverted in sign relative to the input signals. The circuit can be made to give a non-inverted output that is truly equal to the sum of the input voltages by simply wiring an additional unity-gain inverting amplifier between the output of the adder stage and the output of the complete circuit, as shown in the non-inverting unity-gain d.c. adder circuit of *Figure 2.12.*

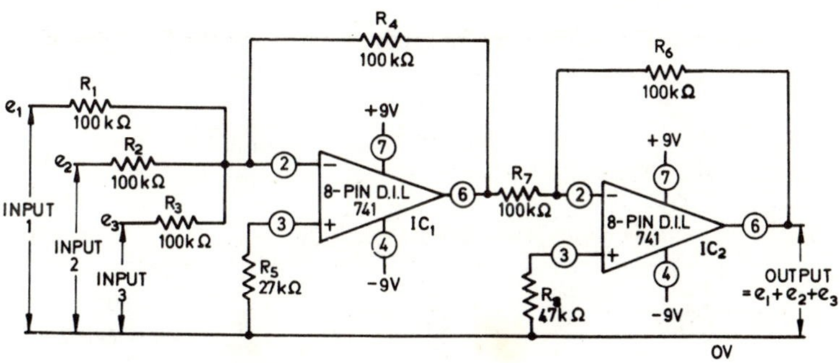

Figure 2.12. Non-inverting unity-gain d.c. adder.

Phase splitter circuits

Pairs of op-amps can be used to make precision balanced phase splitters by wiring the individual op-amps as unity-gain inverting amplifiers, and connecting the two amplifiers in series, as shown in *Figure 2.13*. Here, the output of IC_1 is connected directly to the input of the IC_2 amplifier stage. Consequently, the output of IC_1 is equal in

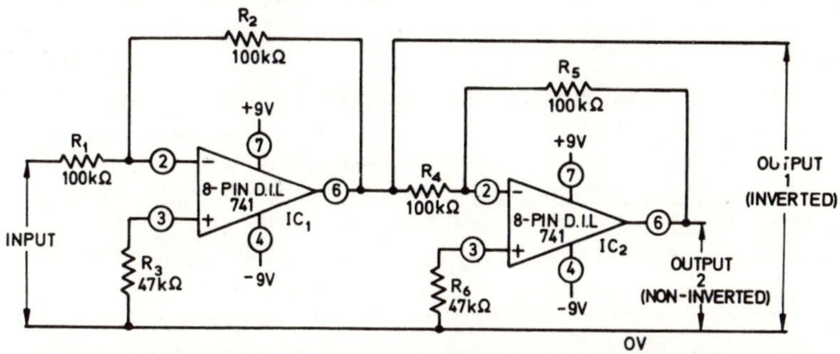

Figure 2.13. Unity-gain balanced d.c. phase-splitter.

amplitude but opposite in phase or polarity to the input signal, and the output of IC_2 is equal in amplitude but opposite in phase to the output of IC_1, and is thus in phase with the input. The two outputs are thus equal in amplitude but opposite in phase relative to each other, and the circuit acts as a unity-gain balanced d.c. phase splitter.

The circuit can be made to give balanced phase-splitting with gain, if required, by simply increasing the gain of the IC_1 inverting amplifier stage. *Figure 2.14* shows the connections for making a variable-gain balanced d.c. phase-splitter. The gain of this circuit can be varied between unity and x 100 via R_3.

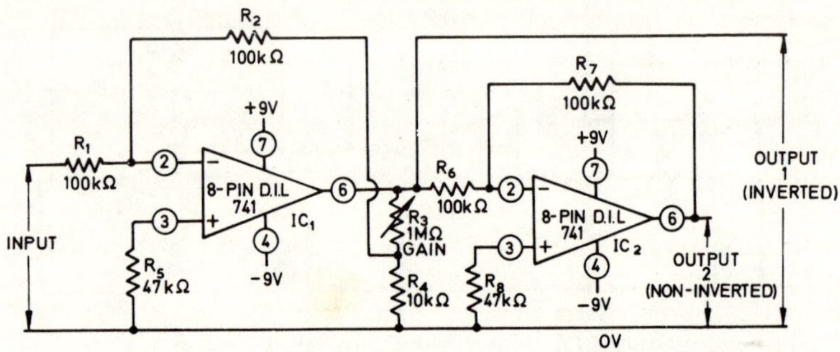

Figure 2.14. Variable-gain (x 1 to x 100) balanced d.c. phase-splitter.

The *Figure 2.13* and *2.14* circuits each have an input resistance of 100 kΩ. In some applications a far greater input resistance than this may be required. In such cases the very-high-impedance variable-gain balanced d.c. phase splitter of *Figure 2.15* can be used. This circuit has an input impedance of hundreds of megohms, and its gain is variable from unity

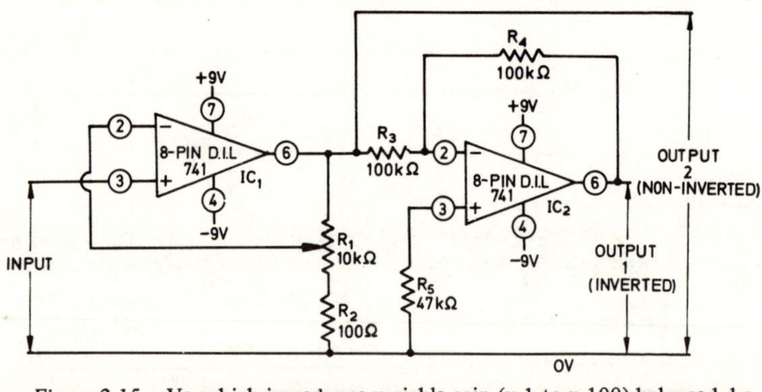

Figure 2.15. Very-high-impedance variable-gain (x 1 to x 100) balanced d.c. phase-splitter.

to x 100 via R_1. IC_1 in this circuit is wired as a variable-gain non-inverting amplifier, with its output feeding directly into the input of the unity-gain inverting stage that is wired around IC_2. The output of IC_1 is in phase with the input signal, and the output of IC_2 is in anti-phase.

Differential amplifiers or subtractors

Operational amplifiers of the 741 type are provided with both inverting and non-inverting input terminals, and can readily be used as differential amplifiers. Differential amplifiers give an output that is proportional to the difference between two input signals, i.e., to the value of one input minus the other, and such circuits are thus capable of carrying out the function of subtraction.

Figure 2.16 shows the practical circuit of a unity-gain differential d.c. amplifier or subtractor. The circuit functions as an inverting amplifier to one input, and as a non-inverting amplifier to the other. Looking first at the input-1 circuitry, it can be seen that if the input-2 terminal is grounded the $R_1 - R_2$ potential divider makes the op-amp work as a x 2 non-inverting amplifier, but the $R_3 - R_4$ potential divider causes only half of the input-1 signal to appear at the positive terminal of the op-amp, so that a non-inverted overall gain of unity takes place between the input-1 terminal and the output.

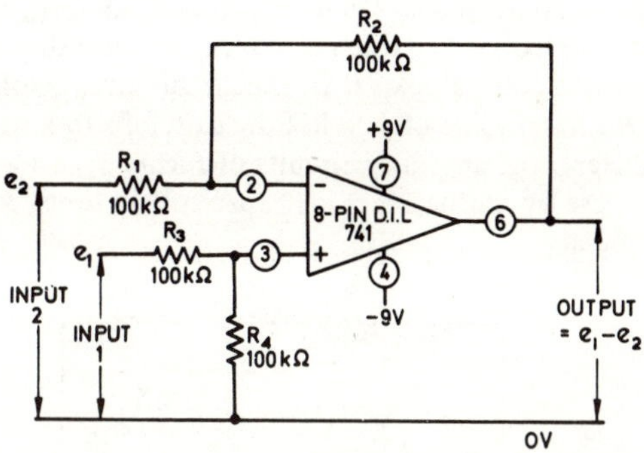

Figure 2.16. Unity-gain differential d.c. amplifier, or subtractor.

Looking now at the input-2 circuitry, it can be seen that the positive terminal is effectively grounded via R_4, and resistors R_1 and R_2 make the op-amp function as a unity-gain inverting amplifier. Thus, the circuit gives unity gain to both inputs, but the input-2 signal gives an inverted output, and the input-1 signal gives a non-inverted output. Consequently, the outputs tend to oppose each other, and the output is equal to input-1 minus input-2. The circuit can be used to carry out the function of subtraction.

The *Figure 2.16* circuit can, if required, be made to give voltage gain by suitably selecting the divider resistor values. The resistors can be given any values on condition that the ratio of R_1 to R_2 is the same as that of R_3 to R_4, in which case the voltage gain is equal to R_2/R_1. *Figure 2.17* shows suitable values for making a x 10 differential d.c. voltage amplifier or subtractor with gain.

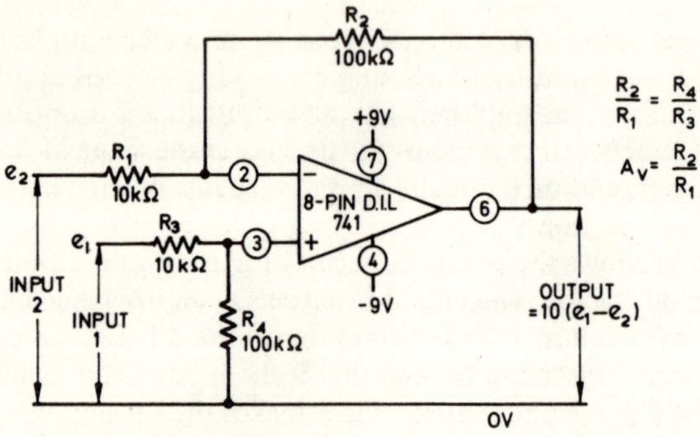

Figure 2.17. x 10 differential d.c. amplifier, or subtractor with gain.

Finally, *Figure 2.18* shows the connections for making a variable-gain differential d.c. amplifier, in which the gain can be varied from x 4 to x 22 via a single variable resistor. Here, resistors R_2 and R_4 are centre-tapped and are coupled via variable resistor R_5 and limiting resistor R_6. When R_5 is adjusted to a value of 0 Ω, the circuit gives a voltage gain of 22, and when R_5 has a value of 100 kΩ the gain falls to x 4.

The three differential amplifier circuits of *Figure 2.16* to *2.18* can be adapted for a.c. use by wiring blocking capacitors in series with each of their input terminals.

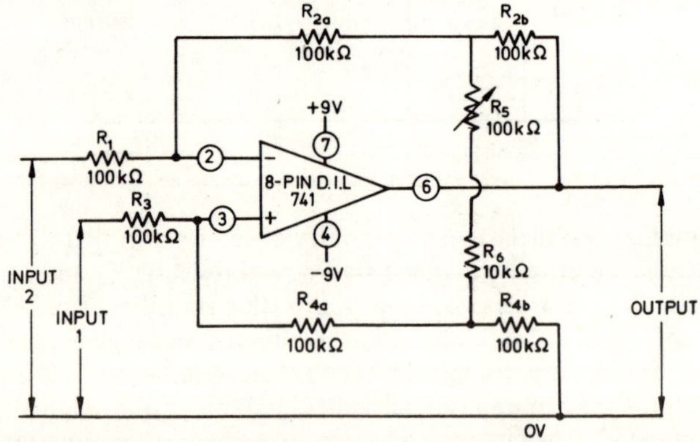

Figure 2.18. Variable-gain (x 4 to x 22) differential d.c. amplifier.

A non-linear (semi-log) amplifier

All the circuits that we have looked at so far in the chapter have been used to give linear voltage amplification, and have used simple resistive feedback elements. Op-amp circuits can be made to give non linear amplification by simply incorporating non-linear elements in their feedback paths. *Figure 2.19* shows a particularly useful type of non-linear amplifier. This circuit in fact gives a semi-log scale of voltage gain. This type of amplification is obtained because silicon diodes D_1 and D_2 are used as the negative feedback elements in an inverting amplifier circuit, and the forward current of a silicone diode varies in approximate proportion to the log of the applied diode voltage. With near-zero applied voltage the diodes act like very high resistances, so the circuit gain is high. With large applied voltages the diodes act like very low resistances, so the circuit gain is low.

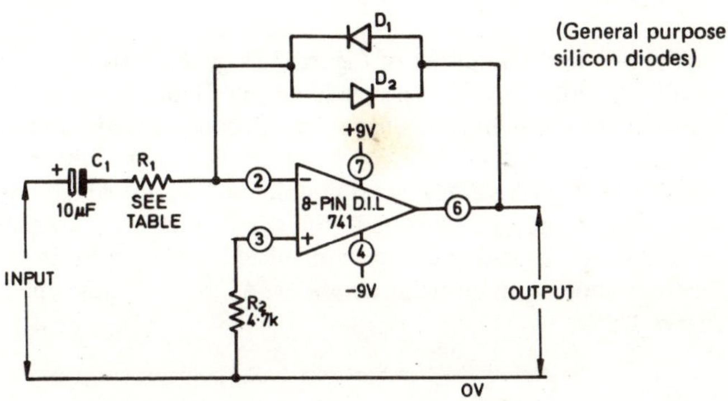

INPUT VOLTS (R.M.S)	R_1 = 1kΩ		R_1 = 10kΩ	
	V_{OUT}(R.M.S)	GAIN	V_{OUT}(R.M.S)	GAIN
1 mV	110 mV	X 110	21 mV	X 21
10mV	330 mV	X33	170mV	X 17
100mV	450 mV	X4·5	360mV	X 3·6
1V	560 mV	X0·56	470mV	X 0·47
10V	600 mV	X0·07	560mV	X0·056

Figure 2.19. Circuit and performance table of non-linear (semi-log) a.c. voltage amplifier.

The table in *Figure 2.19* shows the measured circuit performance of the prototype amplifier when using two alternative values of input resistance. Using a 1 kΩ value of input resistance the circuit gives an r.m.s. output of 600 mV with a 10 V input, and a 330 mV output with an input of 10 mV, i.e., a 1 000:1 change in input causes only a 2:1 change in output. The range of compression can be adjusted by using alternative values of R_1. It should be noted that this circuit gives an approximately square wave output when fed with a sine wave input.

This non-linear type of amplifier is particularly useful as an a.c. bridge-balance detector, in which case the output of the amplifier should be taken to the 1 V or 300 mV range of an a.c. millivolt-meter. The output of an a.c. measuring bridge varies over very wide limits between the balanced and unbalanced states, and it is necessary for the operator to frequently adjust the level of the output-level control in most instruments. If the bridge output is taken to the non-linear amplifier, however, this sensitivity adjustment can be eliminated, since the unit enables voltage level variations over a range of about 10 000:1 to be accommodated on a single range of a millivoltmeter.

Constant-volume amplifier circuits

The non-linear amplifier of *Figure 2.19* gives a virtually constant-amplitude output signal over a wide range of input signal levels, but achieves this constant-level output by introducing heavy amplitude distortion to the signal. In many applications it is desirable to have a circuit which gives a constant-amplitude output, but which does so without introducing distortion to the signal. Such a circuit can be built by using a self-adjusting voltage-controlled linear element in the negative feedback loop of an inverting amplifier. A circuit of this type is shown in *Figure 2.20a*.

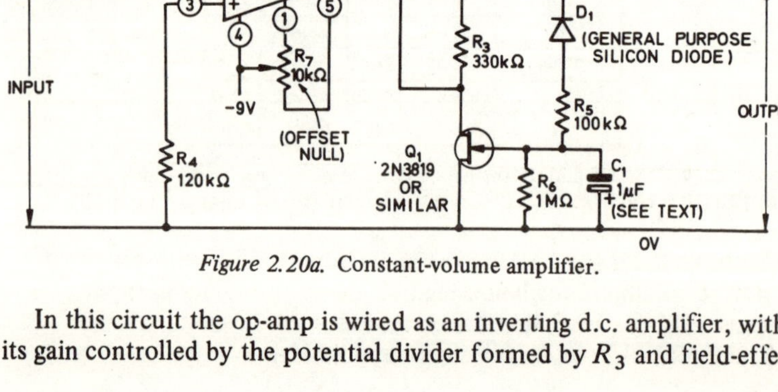

Figure 2.20a. Constant-volume amplifier.

In this circuit the op-amp is wired as an inverting d.c. amplifier, with its gain controlled by the potential divider formed by R_3 and field-effect transitor Q_1. In this particular application the f.e.t. is used as a voltage-controlled resistor, the control bias voltage being obtained from the op-amp output via $D_1 - R_5 - R_6$ and C_1. With zero bias applied to Q_1 gate the f.e.t. acts like a resistance of a few hundred ohms; with a large negative bias applied to the gate the f.e.t. acts like an open circuit.

Thus, when a low-amplitude signal is applied to the op-amp input a small signal voltage tends to appear at the op-amp output. Under this condition very little negative bias is developed at the f.e.t. gate, so the f.e.t. acts like a resistance of only a few hundred ohms. The potential divider action of R_3 and Q_1 results in very little negative feedback under this condition, so the circuit gives a very high voltage gain and tends to increase the op-amp output signal to a reasonable level.

When a large-amplitude signal is applied to the op-amp input, on the other hand, a large signal voltage tends to appear at the op-amp output. Under this condition a large negative bias is developed at the f.e.t. gate, so the f.e.t. acts like an open circuit. Negligible voltage divider action takes place between R_3 and Q_1 under this condition, so heavy negative feedback is applied to the op-amp via R_3, and the circuit gives a very low voltage gain and tends to reduce the op-amp output signal to a reasonable level. Self-regulation of the signal output level thus takes place, and does so without introducing appreciable distortion to the signal.

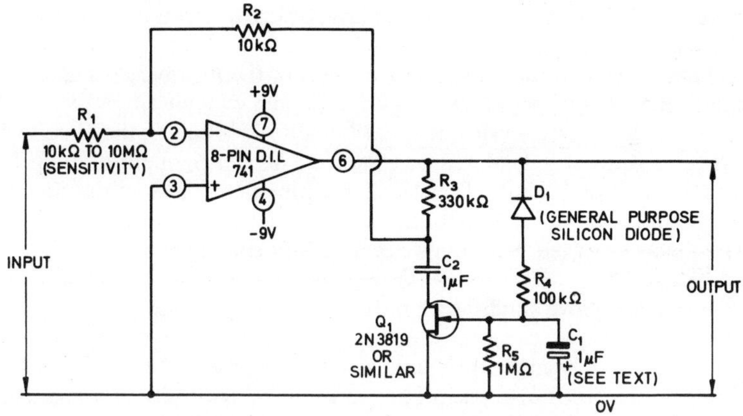

Figure 2.20b. Improved constant-volume amplifier.

In practice the *Figure 2.20a* circuit is capable of giving a virtually constant-amplitude output signal over a 30 dB range of input signal levels, the actual signal operating range being determined by R_1. R_1 is selected to handle the *maximum* input signal required, since the output becomes distorted when this level is exceeded.

A minor snag of the *Figure 2.20a* circuit is that it has an inherently poor d.c. stability, and requires the use of offset null control R_7 to compensate for this defect. The poor d.c. stability occurs because negligible d.c. negative feedback takes place under low-level input conditions, when the f.e.t. is acting as a low resistance and the circuit is giving a very high gain. This snag can be overcome, and the need for an offset null control eliminated, by wiring a blocking capacitor between the drain of Q_1 and the $R_2 - R_3$ junction, as shown in the improved constant-volume amplifier circuit of *Figure 2.20b.* The capacitor acts as an open circuit to d.c., so a high degree of d.c. negative feedback is applied to the op-amp via R_2 and R_3, and the circuit has good d.c. stability. The capacitor acts as a short circuit to a.c. at normal operating frequencies, however, so the voltage divider action of R_3 and Q_1 is unimpaired and the circuit acts as a constant-volume amplifier to a.c. signals.

The *Figure 2.20a* and *2.20b* circuits give virtually identical performances. In both cases R_1 determines the signal operating range of the circuit, and is selected to suit the *maximum* input signal that the circuit is expected to handle. The R_1 value is selected on the basis of 200 kΩ/V of r.m.s. input signal: for a maximum input of 50 V R_1 is given a value of 10 MΩ, and for a maximum input of 50 mV R_1 is given a value of 10 kΩ. Capacitor C_1 determines the 'follow' or a.g.c. time constant of the circuit, and its value can be changed to suit individual needs: reducing the C_1 value reduces the time constant, and increasing C_1 increases the time constant.

Table 2.1 shows the typical performance of the *Figure 2.20 a* and *2.20b* circuits at different input signal levels and with alternative R_1 values. With R_1 given a value of 1 MΩ (to suit a 5 V input), the circuit gives an output of 2.85 V with an input of 5 V, and an ouput of 1.48 V with an input of 100 mV. The effective compression range of the circuit is roughly 30 dB. Two or more of these constant-volume circuits can be wired in series, if required, to give even greater compression.

Frequency-selective amplifier circuits

Op-amps can be made to function as frequency-selective amplifiers by wiring reactive resistor-capacitor networks into their feedback loops. They can be made to act as frequency-selective tuned amplifiers or acceptor filters, as notch or rejector filters, or as high pass or low-pass amplifiers, etc. Five useful frequency-selective circuits are shown in this final section of this chapter.

Figure 2.21 shows the practical circuit of a 1 kHz frequency-selective tuned amplifier or acceptor filter. The circuit exhibits characteristics similar to those of an *LC* tuned amplifier with a *Q* of about 50. The circuit gives a gain of x 200 to signals at the centre frequency of 1 kHz,

Table 2.1 Performance results for the *Figure 2.20a* and *b* circuits.

V_{IN} $(R_1 = 10K\Omega)$	V_{IN} $(R_1 = 100K\Omega)$	V_{IN} $(R_1 = 1M\Omega)$	V_{IN} $(R_1 = 10M\Omega)$	V_{OUT}
50 mV	500 mV	5 V	50 V	2.85 V
20 mV	200 mV	2 V	20 V	2.81 V
10 mV	100 mV	1 V	10 V	2.79 V
5 mV	50 mV	500 mV	5 V	2.60 V
2 mV	20 mV	200 mV	2 V	2.03 V
1 mV	10 mV	100 mV	1 V	1.48 V
500 μV	5 mV	50 mV	500 mV	0.89 V
200 μV	2 mV	20 mV	200 mV	0.40 V
100 μV	1 mV	10 mV	100 mV	0.20 V
50 μV	500 μV	5 mV	50 mV	0.10 V

but at 500 Hz (one octave down) the gain is roughly x 3, and at 2 kHz (one octave up) the gain is roughly x 2. The gain falls to unity at 150 Hz and 3.3 kHz.

The operating theory of the *Figure 2.21* circuit is quite simple. The op-amp is wired as an inverting amplifier with twin-T filter $R_2 - R_3 - R_4$ and $C_2 - C_3 - C_4$ wired in the negative feedback loop between the output and the negative input terminal. In this application the twin-T filter acts like a frequency-controlled resistor that presents a near-infinite

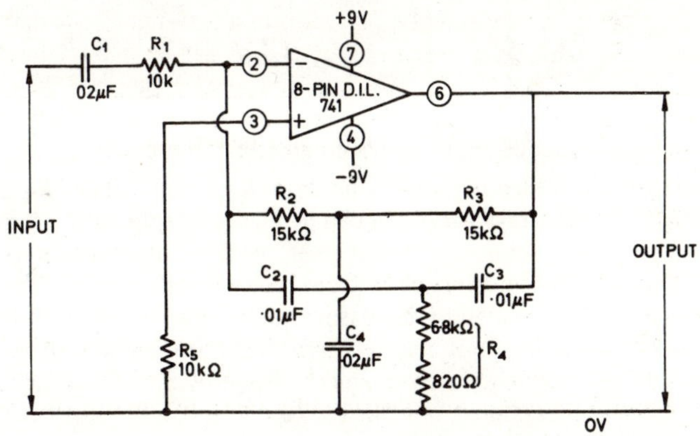

Figure 2.21. 1 kHz tuned (acceptor) amplifier (twin-T).

impedance at the centre frequency, but a low impedance at all other frequencies. Consequently, at the centre frequency negligible negative feedback is applied to the circuit, and very high gain is available, but at all other frequencies heavy negative feedback is applied, and the gain is low.

The centre frequency of the circuit is determined by the twin-T component values, and these can be changed to satisfy individual requirements. The twin-T resistors should ideally be kept in the ratios $R_2 = R_3 = 2 \times R_4$, and the twin-T capacitors must be kept in the ratios $C_2 = C_3 = C_4/2$, in which case the centre frequency of the circuit $= 1/6.28\, R_2.C_2$. In practice the stability of the circuit is enhanced by making R_2 and R_3 very slightly greater than $2 \times R_4$ (by about 2 %). The C_1 input capacitor value of the design is selected to improve the low-frequency rejection of the circuit, and should be given the same value as C_4.

Figure 2.22 shows how the *Figure 2.21* circuit can be changed into a 1 kHz notch or rejector circuit by repositioning the twin-T filter. This circuit totally rejects input signals at the centre frequency of 1 kHz, but accepts and gives unity gain to all other input signals. The centre

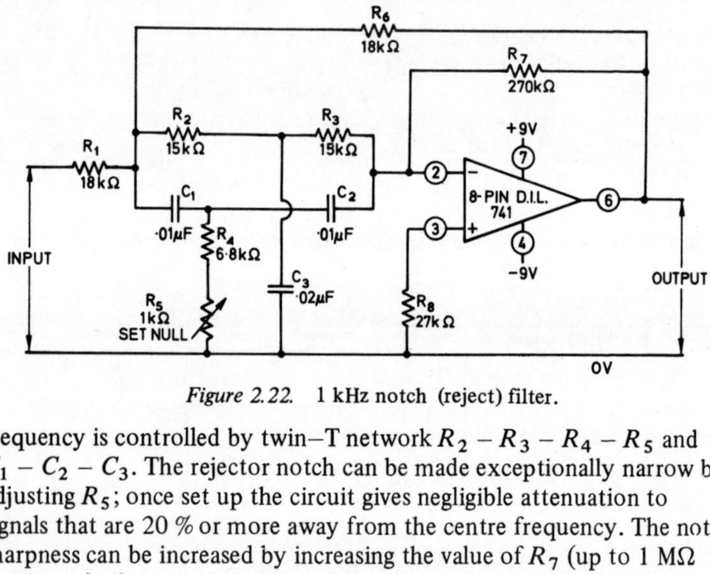

Figure 2.22. 1 kHz notch (reject) filter.

frequency is controlled by twin–T network $R_2 - R_3 - R_4 - R_5$ and $C_1 - C_2 - C_3$. The rejector notch can be made exceptionally narrow by adjusting R_5; once set up the circuit gives negligible attenuation to signals that are 20 % or more away from the centre frequency. The notch sharpness can be increased by increasing the value of R_7 (up to 1 MΩ maximum), if required, but a perfect null then becomes more difficult to obtain. The off-frequency gain can be increased by increasing the value of R_6 (to 1.8 MΩ maximum), but this increase in gain is obtained at the expense of reduced notch sharpness.

The two frequency-selective circuits that we have looked at so far have both used twin-T filters as their frequency-selecting elements. Other types of *RC* frequency-selecting networks can also be used in op-amp circuits. *Figure 2.23*, for example, shows how a Wien network

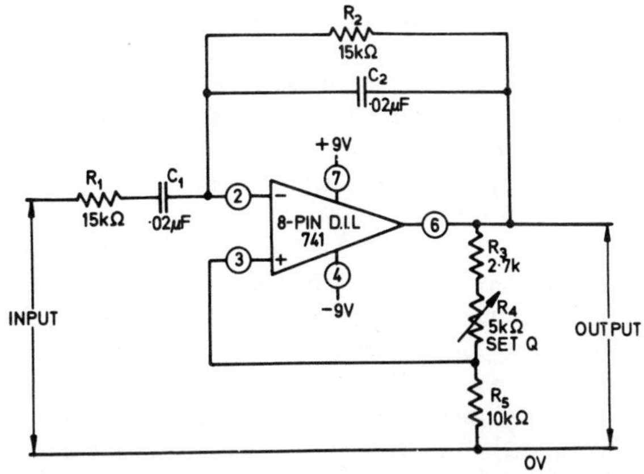

Figure 2.23. 1 kHz tuned (acceptor) amplifier (Wien).

$(R_1 - R_2 - C_1 - C_2)$ can be connected to an op-amp so that it functions as a 1 kHz frequency-selective tuned amplifier or acceptor filter.

This circuit employs two feedback paths. One path is from the output to the negative input terminal of the op-amp via the Wien network, and the other is from the output to the positive input terminal via the $R_3 - R_4$ and R_5 potential divider. The circuit action is such that the two feedback signals cancel each other out at the Wien centre frequency, so the circuit gives very high gain to centre-frequency signals. At all other frequencies the two signals do not self-cancel, so the circuit gives low gain to all off-tune frequencies.

Ideally, R_5 should have a value exactly double the sum of R_3 and R_4; slight errors in these ratios result in either oscillation (R_5 value too large) or very low effective Q (R_5 value too small). R_4 is used to set the circuit Q or tuning sharpness to an acceptable level, consistent with good circuit stability. The centre frequency of the circuit can be made variable by replacing R_1 and R_2 with a two-gang variable resistor.

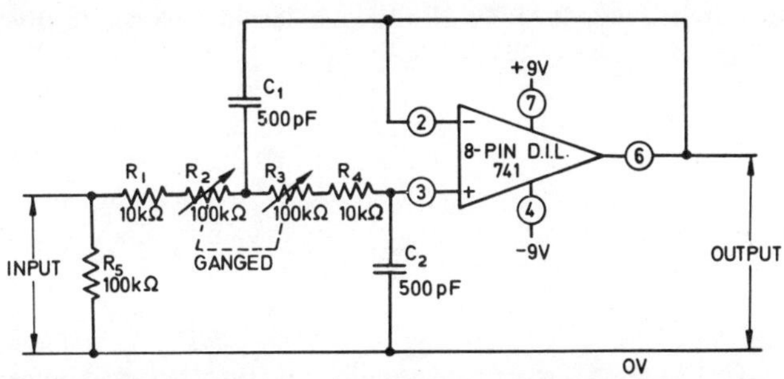

Figure 2.24. Variable low-pass filter, covering 2.2 kHz to 24 kHz.

Figure 2.24 shows the circuit of a variable low-pass filter. This circuit gives unity gain to all frequencies below a certain cut-off value, but attenuates all signals above the cut-off frequency. The cut-off frequency is defined here as the frequency at which the output signal falls by 3 dB relative to the input signal. The cut-off frequency of the circuit can be varied via the two-gang $R_2 - R_3$ variable resistor from 2.2 kHz ($R_2 = R_3 = 100$ kΩ) to 24 kHz ($R_2 = R_3 = 0$ Ω). Beyond the cut-off frequency the circuit exhibits a 2nd order response in which the gain falls by 12 dB for each octave increase in frequency, or by 40 dB for each decade increase in frequency.

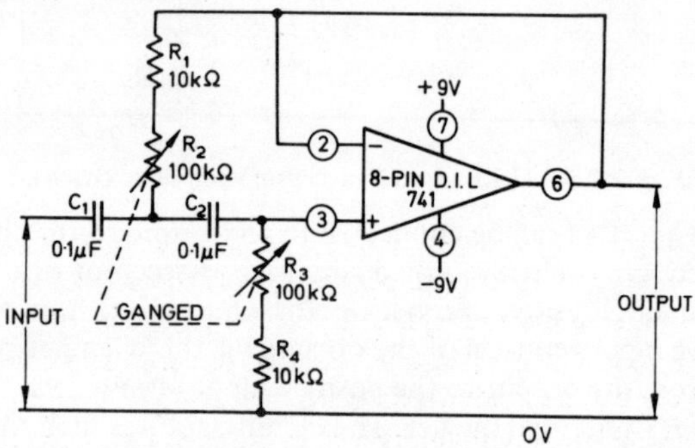

Figure 2.25. Variable high-pass filter, covering 235 Hz to 2.8 kHz.

Finally, *Figure 2.25* shows the circuit of a variable high-pass filter. This circuit gives unity gain to all frequencies above a certain cut-off value, but attenuates all signals below the cut-off frequency. The cut-off frequency is again defined as the frequency at which the output signal falls by 3 dB relative to the input signal, and can be varied via R_2 and R_3. The cut-off frequency is variable from 235 Hz ($R_2 = R_3 = 100$ kΩ) to 2.8 kHz ($R_2 = R_3 = 0$ Ω). Beyond the cut-off point the circuit has a 2nd order response in which the gain falls by 12 dB for each octave decrease in frequency, or by 40 dB for each decade decrease in frequency.

CHAPTER 3

25 INSTRUMENTATION PROJECTS

Operational amplifiers can be used in a variety of instrumentation circuit applications. They can be used as precision voltage sources, as variable voltage supplies, as stabilised power supply units, and as precision rectifiers and a.c./d.c. converters. When used in conjunction with moving coil meters they can be made to function as d.c. and a.c. voltmeters and millivoltmeters, as d.c. microammeters, as linear-scale ohmmeters, and as linear-scale capacitance meters.

Twenty-five useful op-amp instrumentation projects of various types are described in the present chapter. All of these projects are designed around the popular type 741 integrated-circuit op-amp, and the pin connections shown in the following diagrams apply to the 8-pin dual-in-line version of this device only.

Precision voltage source circuits

Electronics laboratories often need a precision voltage reference or source for calibrating instruments accurately. A Weston standard cell generates an accurate potential of 1.018 V, and is useful as a voltage standard. These cells normally have an output impedance of one or two kilohms, however, and this relatively high output impedance makes the devices unsuitable for use as precision voltage sources at output currents in excess of one microamp or so. *Figure 3.1* shows a circuit that overcomes the basic disadvantage of the Weston cell, and enables it to be used as a precision voltage source at currents up to several milliamps.

The op-amp in the *Figure 3.1* circuit is connected as a unity-gain voltage follower, with the Weston cell connected directly to its positive input terminal. The op-amp has a very high input impedance in this application, and typically draws only 0.03 μA from the Weston cell, but has an output impedance of virtually zero and can provide output currents up to 5 mA or so. Thus, the circuit gives a precision output of 1.018 V, and can supply currents up to about 5 mA.

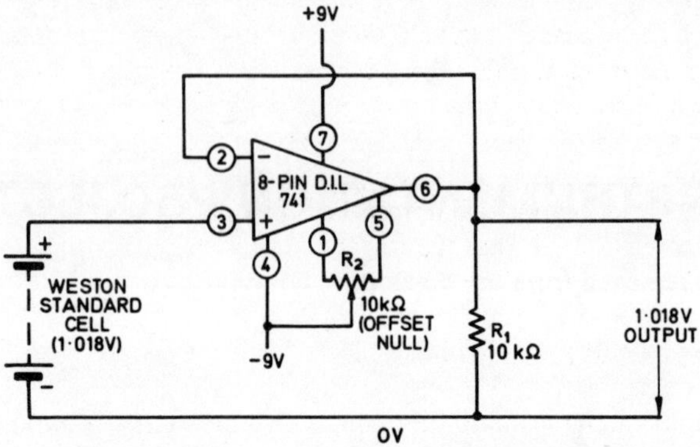

Figure 3.1. Precision voltage source.

If required, the output current capability of the circuit can be boosted by wiring an emitter follower stage in series with the op-amp output, as shown in the precision voltage source circuit of *Figure 3.2*. Note that the base-emitter junction of the emitter follower is in series with the op-amps negative feedback loop, so the junction does not inhibit the voltage following action of the circuit.

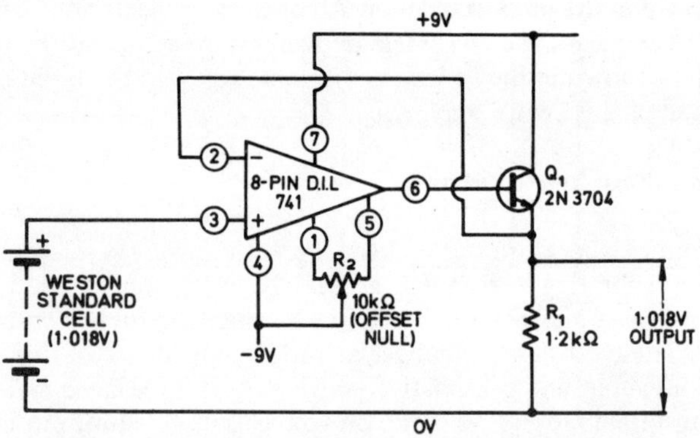

Figure 3.2. Precision voltage source with boosted output.

Also note that a 10 kΩ offset null control is used in both of these precision voltage source circuits. This control is used to ensure precise tracking of the input and output voltages, and is simply set to give zero voltage differences between the positive input and the output terminal of the op-amp with the Weston cell connected in place. If output voltage errors of a few millivolts can be tolerated, the offset null control can be eliminated from the circuit.

Normally, the *Figure 3.1* or *3.2* circuits should be built into a small box, complete with battery power supply and on-off switch, but with the Weston cell connected to the circuit externally via a pair of input terminals. In this case the cell should not be connected to the circuit until after the on-off switch controlling the op-amp has been turned on. If preferred, however, the Weston cell can be enclosed in a box with the rest of the circuit, but in this case a switch should be wired in series between the Weston cell and the positive terminal of the op-amp and ganged to the main on-off switch, so that the cell is automatically disconnected when power is removed from the op-amp.

Variable power supply circuits

Op-amps can be made to function as high-performance variable power supply circuits in a number of alternative ways. One simple way of making an op-amp function as a variable power supply is shown in *Figure 3.3*. This circuit gives an output that is fully adjustable from 0V to 12 V at currents up to a maximum of about 50 mA.

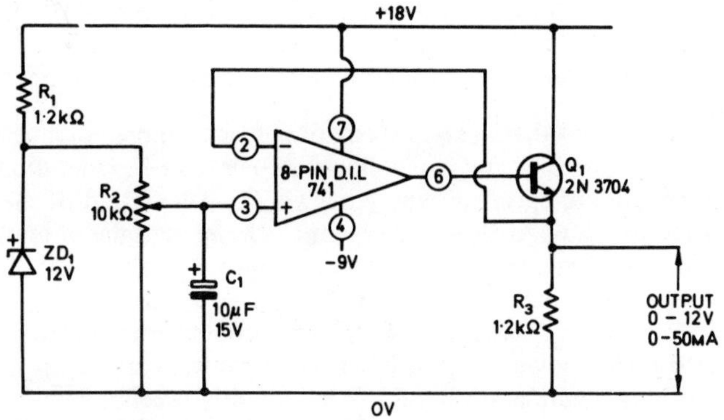

Figure 3.3. Simple variable-voltage supply.

The operation of the *Figure 3.3* circuit is quite simple. ZD_1 is a zener diode, and is energised from the positive supply line via R_1. A constant reference potential of 12 V is developed across the zener diode, and this voltage is fed to variable potential divider R_2. The output of the potential divider is fully variable from 0V to 12 V, and is fed to the positive input terminal of the op-amp. The op-amp is wired as a unity-gain voltage follower, with Q_1 connected as an emitter follower current-booster stage in series with the output. Thus, the output voltage of the circuit follows the voltage set at the positive terminal via R_2, and is fully variable from

0V to 12 V. Note that this particular circuit uses an 18 V positive supply and a 9 V negative supply.

An alternative type of variable power supply circuit is shown in *Figure 3.4*. The output of this circuit is fully variable from + 3 V to + 15 V, at currents up to a maximum of about 50 mA. Although the output of the circuit is not fully adjustable down to 0 V, the circuit has the great advantage of using only a single 18 V positive supply.

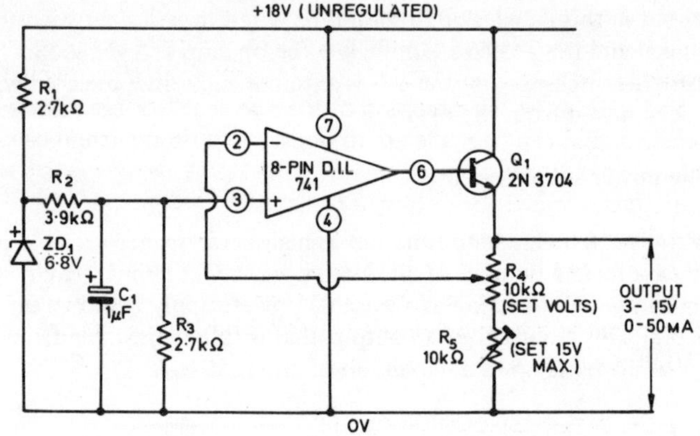

Figure 3.4. Simple 3 V - 15 V variable supply.

The operation of the *Figure 3.4* circuit is slightly more complicated than that of *Figure 3.3*, and is as follows. ZD_1 is a 6.8 V zener diode, and is energised from the positive supply line via R_1. The output of the zener diode is fed to the fixed $R_2 - R_3$ potential divider, and the potential divider applies a fixed reference voltage of approximately 3 V to the positive input terminal of the op-amp. The op-amp is wired as a variable-gain non-inverting d.c. amplifier, with Q_1 connected as an emitter follower current-booster stage in series with its output, and with variable gain-determing potential divider $R_4 - R_5$ wired between the emitter follower output and ground.

With the slider of R_4 set to the Q_1 emitter position, the amplifier operates with 100 % negative feedback, and thus gives unity-gain: under this condition the amplifier acts as a voltage follower, and gives an output of 3 V, i.e., the same potential as the reference voltage that is applied to its positive input terminal. With the slider of R_4 set to the $R_4 - R_5$ junction, the amplifier operates with reduced negative feedback, and gives a voltage gain of approximately x 5: under this condition the circuit gives an output of 15 V, i.e., 5 times the 3 V reference voltage. In practice R_5 is pre-set so that the circuit gives a maximum output of 15 V, and the output is then fully variable via R_4 from 3 V to 15 V at currents up to 50 mA. Since the gain-determing $R_4 - R_5$ potential divider is fed directly from the units output terminal, the output voltage of the circuit is fully stabilised and is virtually unaffected by variations in output load current.

A number of useful modifications and improvements can be made to the simple variable supply circuit of *Figure 3.4*. The regulation of the output voltage, for example, depends on the precision of the reference voltage that is applied to the positive input terminal of the op-amp. In the *Figure 3.4* circuit the reference voltage is simply derived from a zener diode that is energised from the unregulated 18 V supply line. One improvement, therefore, is to feed the zener diode from a pre-regulated supply, and so improve the precision of the reference voltage. Other improvements that can be made are to boost the output current capability of the circuit by replacing Q_1 with a super-alpha-connected pair of power transistors, and to boost the output voltage range of the circuit by using a higher voltage supply.

Figure 3.5 shows the practical circuit of a variable power supply that incorporates all three of these improvements and modifications. The circuit is capable of providing output voltages that are fully variable from 3 V to 30 V at currents up to 1 A.

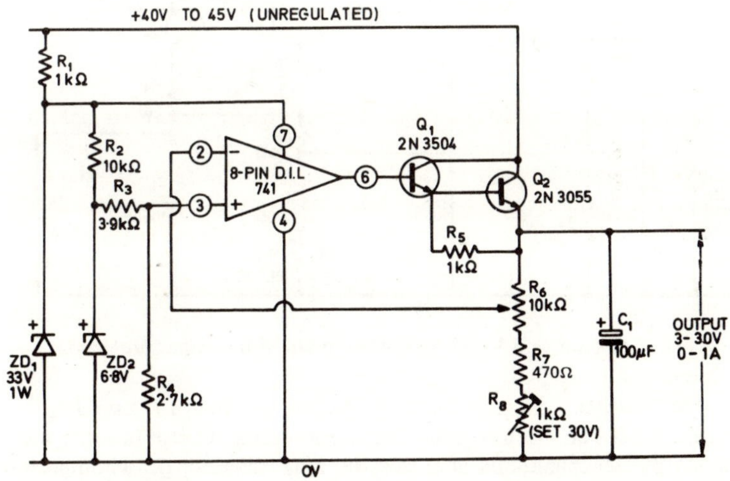

Figure 3.5. 3 V - 30 V, 0-1 amp stabilised p.s.u.

The *Figure 3.5* circuit is powered from a 40 V to 45 V unregulated supply. This supply is fed directly to the collectors of output transistors Q_1 and Q_2, but is fed indirectly to the op-amp via R_1 and zener diode ZD_1. The pre-regulated 33 V output of ZD_1 is also used to feed reference zener diode ZD_2 and thence to provide a highly stable reference potential of 3 V to the positive input terminal of the op-amp via potential divider $R_3 - R_4$. Transistors Q_1 and Q_2 are super-alpha connected in series with the output of the op-amp, and the transitor/op-amp combination is connected as a variable-gain non-inverting d.c. amplifier, with gain control via variable potential divider $R_6 - R_7 - R_8$. The potential divider enables the gain of the amplifier to be varied between unity and x 10. The output voltage of the circuit can thus be varied between 3 V and 30 V, at currents up to 1 A, and is fully stabilised.

A major weakness of the *Figure 3.5* circuit is that it has no short-circuit protection, and the circuitry may be damaged if a short occurs across the output terminals. One way around this snag is to fit a fuse in series with the output, but a far better solution is to fit the circuit with electronic short-circuit protection, as shown in the stabilised power supply circuit of *Figure 3.6*.

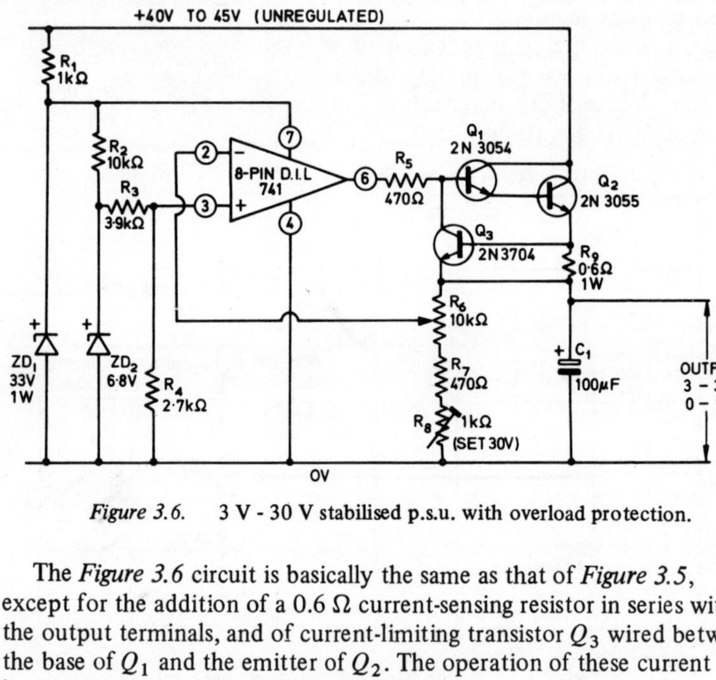

Figure 3.6. 3 V - 30 V stabilised p.s.u. with overload protection.

The *Figure 3.6* circuit is basically the same as that of *Figure 3.5*, except for the addition of a 0.6 Ω current-sensing resistor in series with the output terminals, and of current-limiting transistor Q_3 wired between the base of Q_1 and the emitter of Q_2. The operation of these current limiting components is quite simple. Q_3 is a silicon transistor, and needs a forward-base emitter voltage of greater than 600 mV to turn on. The base-emitter voltage of the transistor is derived from 0.6 Ω resistor R_9 in series with the output terminals, and the magnitude of this voltage is dictated by the output current of the circuit.

Normally, the output current of the circuit is less than 1 A, so insufficient voltage is developed across R_9 to bias Q_1 on: under this condition Q_3 acts like an open circuit, and has no effect on the operation of the system. If a short is placed across the circuits output, on the other hand, the output current will tend to rise above 1 A, and at least 600 mV will be developed across R_9: under this condition Q_3 is biased on, and acts as a shunt between the base of Q_1 and the emitter of Q_2, and tends to turn Q_1 and Q_2 off and so reduce the output current of the circuit. In practice, a large degree of degeneration tends to take place via the $Q_1 - Q_2 - Q_3 - R_9$ loop, and the output current automatically self-limits at a value of 1 A if a short occurs at the output. The *Figure 3.6* circuit thus provides an output that is fully variable from 3 V to 30 V at currents up to 1 A, but has fully automatic overload protection and will not be damaged by short-circuits at the output terminals.

Each of the four variable power supply circuits that we have looked at so far has a single pair of output terminals, and acts as a single power source. In cases where two sets of supplies are required (as in the case of op-amp circuits), two separate single-output power supply units must be interconnected to provide the necessary supplies. A possible alternative, that can be used in cases where the two supplies are required to be of equal amplitude but opposite polarity, is to use a single power supply with a centre-tapped output. The circuit of a simple power supply of this type is shown in *Figure 3.7*.

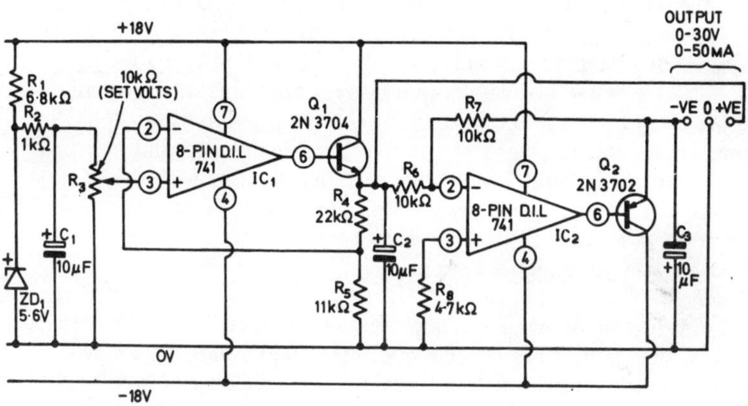

Figure 3.7. Simple centre-tapped 0-30 V p.s.u., for driving op-amp circuits.

The *Figure 3.7* circuit has three output terminals, marked '-ve', '0', and '+ve', and can provide three sets of outputs. The three outputs can be varied simultaneously from 0V to +15 V between the 0 and +ve terminals, from 0V to −15 V between the 0 and -ve terminals, and between 0V and +30 V between the -ve and +ve terminals. The circuit can provide maximum output currents of roughly 50 mA between any set of terminals. The circuit operates as follows.

Zener diode ZD_1 is used to generate a fixed reference potential of 5.6 V from the 18 V positive supply line via R_1. This reference voltage is fed to variable potential divider R_3 via R_2 and C_1, making a variable potential of 0 V to +5 V available at the R_3 slider. This variable reference potential is fed to the positive input terminal of IC_1, which (together with npn transistor Q_1) is wired as a x 3 non-inverting d.c. amplifier, with its output taken to the +ve output terminal of the power supply circuit. Thus, the potential between the 0 and +ve output terminals can be varied between 0V and +15 V via R_3, and output currents up to 50 mA are available.

IC_2 and pnp transistor Q_2 of the design are wired together as a unity-gain inverting d.c. amplifier, with input applied from the +ve output terminal of the power unit, and the output taken to the units −ve output terminal. Consequently, the output potential between the 0 and −ve terminals is equal in magnitude but opposing in polarity to that between the 0 and +ve terminals, and can be varied between 0 V and − 15 V via R_3, at output currents up to 50 mA. Since the output at the +ve terminal can be varied between 0 V and + 15V, and that at the −ve terminal can be varied between 0 V and −15 V, it follows that the potential between the −ve and +ve terminals can be varied between 0 V and 30 V via R_3, and that the 0 terminal acts as a centre-tap between this output.

The basic centre-tapped power supply circuit of *Figure 3.7* is capable of supplying maximum output currents of only 50 mA, and incorporates no overload protection. The output current capability of the circuit can be increased to 1 A or so by replacing Q_1 and Q_2 with super-alpha connected pairs of power transistors of suitable polarity, and overload protection can be built in by fitting each output stage with the kind of automatic current limiting described in the *Figure 3.6* circuit.

Precision half-wave rectifier circuits

Conventional diodes act as imperfect rectifiers to low-level a.c. signals, because they do not begin to conduct significantly until the applied signal voltage exceeds a certain 'knee' value. In silicon diodes this knee value is of the order of 600 mV, so silicon diodes give negligible rectification to signal voltages below this level.

Operational amplifiers can be combined with silicon diodes in such a way that the effective knee voltage of the diode is reduced by a factor equal to the open-loop gain of the op-amp, the combination then acting as a near-perfect rectifier even to signals with amplitudes of only a fraction of a millivolt. *Figure 3.8a* shows the practical version of a precision unity-gain half-wave rectifier circuit of this type.

The operation of the *Figure 3.8a* circuit is fairly simple. The op-amp is wired as an inverting amplifier, with input applied via R_3, and negative feedback applied via $R_1 - D_1$ or $R_2 - D_2$. On positive input half-cycles the op-amp output swings negative, so D_2 conducts via R_2, and D_1 is reverse biased: zero output is available at the circuits ouput at the $R_1 - D_1$ junction under this condition. On negative half-cycles the op-amp output goes positive, so D_1 conducts via R_1, and D_2 is reverse biased: a positive output is available at the $R_1 - D_1$ junction under this condition. The circuit thus acts as a half-wave rectifier, and gives a positive output at the $R_1 - D_1$ junction, or a negative output at the $R_2 - D_2$ junction.

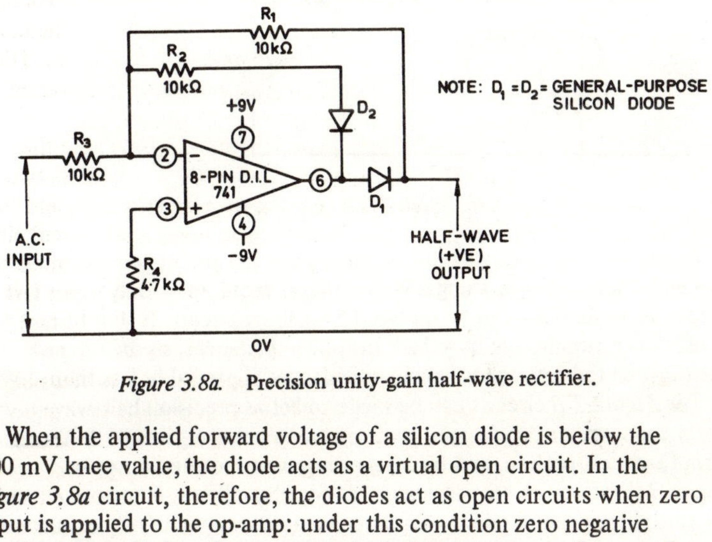

Figure 3.8a. Precision unity-gain half-wave rectifier.

When the applied forward voltage of a silicon diode is below the 600 mV knee value, the diode acts as a virtual open circuit. In the *Figure 3.8a* circuit, therefore, the diodes act as open circuits when zero input is applied to the op-amp: under this condition zero negative feedback is applied to the op-amp, which thus operates in the open-loop mode and gives a voltage gain of about 100 000. Consequently, an input signal of only 6 μV or so is needed to raise the forward diode voltage to the 600 mV knee value beyond which effective rectification takes place, so the circuit effectively reduces the knee voltage by an amount equal to the open loop gain of the op-amp. Once the forward voltage of the diode has exceeded the knee value, the diode acts as a virtual short circuit, and under this condition the gain of the circuit is dictated almost entirely by the $R_1 - R_3$ or $R_2 - R_3$ values. The *Figure 3.8a* circuit gives unity voltage gain under this condition, and acts as a near-perfect half-wave rectifier that gives a positive output from the $R_1 - D_1$ junction, or a negative output from the $R_2 - D_2$ junction.

The overall voltage gain of the *Figure 3.8a* circuit is dictated by the ratios of R_1 to R_3, or R_2 to R_3, as in the case of a conventional inverting amplifier. Normally $R_1 = R_2$, and the voltage gain $A_v = R_1/R_3$. The circuit can thus be made to give voltage gain as well as rectification by simply making the values of R_1 and R_2 larger than that of R_3, as in the case of the precision x 10 half-wave rectifier circuit shown in *Figure 3.8b*.

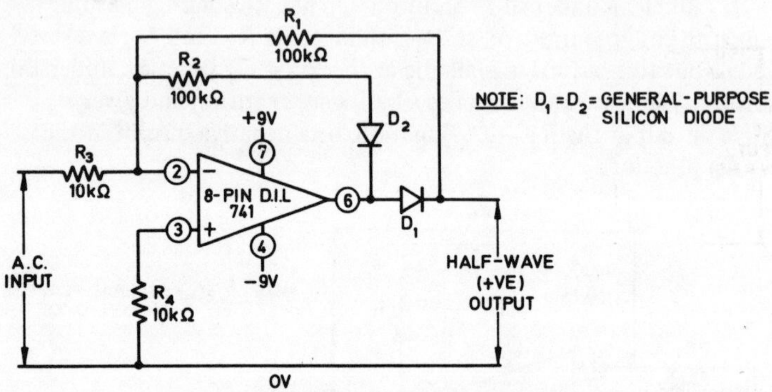

Figure 3.8b. **Precision x 10 half-wave rectifier.**

It should be noted that both of the *Figure 3.8* circuits are designed to give either a positive or a negative output, as required; if only a positive output is needed, R_2 can be replaced by a short circuit. It should also be noted that both circuits have high output impedances, so the outputs must not be fed into loads having impedances appreciably less than 1 MΩ.

The *Figure 3.8* circuits can be made to act as precision half-wave a.c./d.c. converters by designing them to give voltage gains of 2.22 to give form-factor correction, and by integrating their rectified outputs. *Figure 3.9* shows the practical version of such a circuit. Resistors R_1 and R_2 ensure that the required gain is attained, and capacitor C_2 carries out the function of integration. The circuit is designed to give unity conversion gain, with a maximum r.m.s. input and d.c. output of 2 V. Output linearity is better than 0.1 % of full scale, i.e., the accuracy is better than $\pm$2 mV of reading.

The *Figure 3.9* circuit is specifically designed to feed into the high-impedance input of a digital panel meter, and has a high output impedance. If the output is to be fed to a low-impedance indicator, such as a moving coil meter, a high-impedance voltage-follower buffer stage must be interposed between the converter output and the input of the indicator. The converter has an input impedance of only 10 kΩ, and a voltage-follower can also be interposed between the input signal and the input of the converter to give a high input impedance to the entire measuring circuit, if required.

Another application of the *Figure 3.8a* circuit is as a peak-voltage

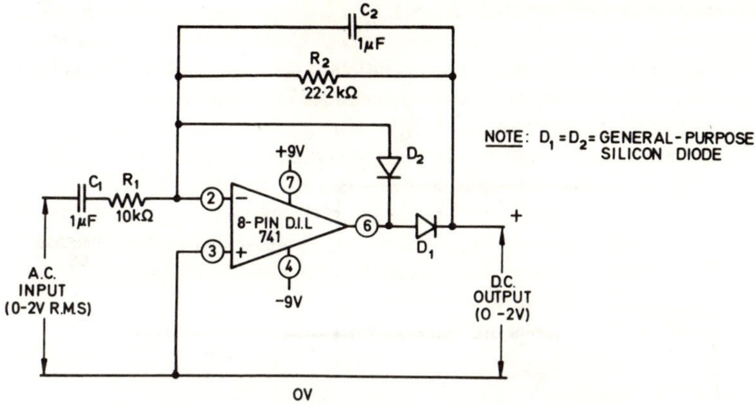

Figure 3.9. Precision half-wave a.c./d.c. converter.

detector: the circuit can be made to read peak voltage by simply connecting a reservoir capacitor across its output terminals. An even simpler peak-voltage detector circuit is shown in *Figure 3.10*.

Here, the op-amp is wired as a unity-gain voltage-follower, with feedback provided via D_1, and with reservoir capacitor C_1 wired between the output and ground, so that C_1 charges to the peak positive value of any input voltage applied to the non-inverting input of the op-amp. The capacitor charges rapidly via D_1 and the low-impedance output of the op-amp, but discharges slowly via the input impedance of the inverting input terminal (typically 1 MΩ). The discharge rate can be increased, if required, by wiring a shunt resistor across the capacitor, as shown. Note that this shunt appears directly across the circuits output, so the circuit has a high effective output impedance.

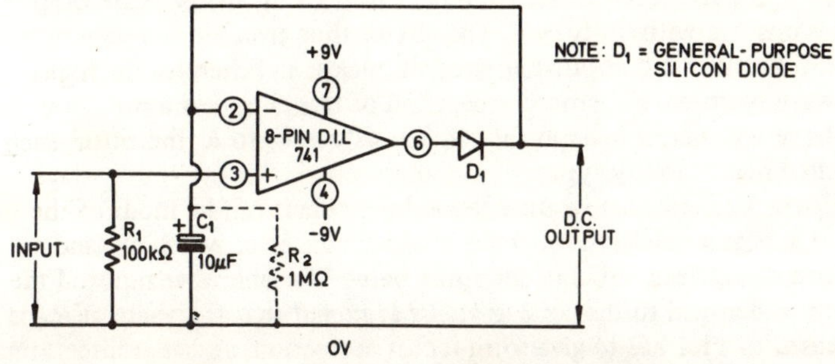

Figure 3.10. Peak voltage detector (positive).

Precision full-wave rectifier circuits

The three op-amp rectifying circuits that we have looked at so far all give half-wave rectification. Op-amps can also be made to function as full-wave rectifiers. *Figure 3.11* shows one method of obtaining full-wave rectification. In this case IC_1 is wired as a precision half-wave rectifier which gives a negative output, and IC_2 is wired as a 2-input inverting adder, with one input taken directly from the circuits input terminal, and the other taken from the output of IC_1. The adder gives a gain of unity to the direct-input signal, but a gain of x 2 to the IC_1 output signal.

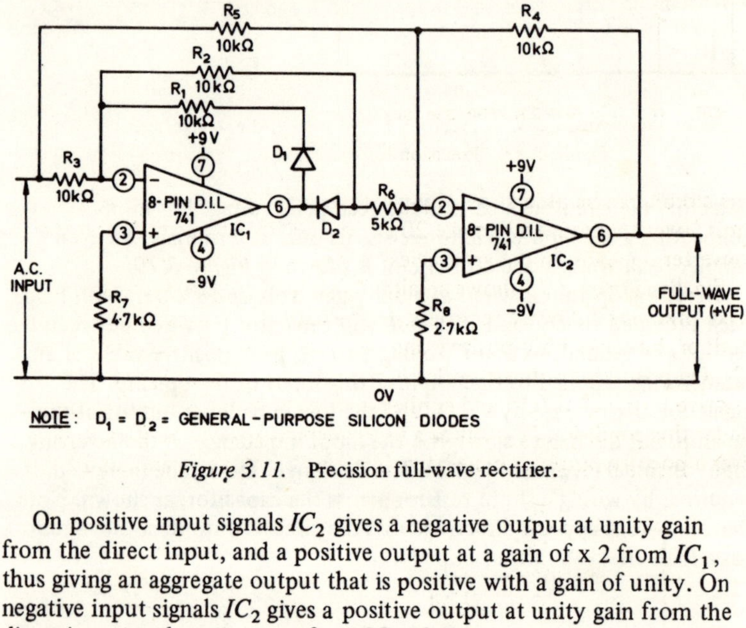

Figure 3.11. Precision full-wave rectifier.

On positive input signals IC_2 gives a negative output at unity gain from the direct input, and a positive output at a gain of x 2 from IC_1, thus giving an aggregate output that is positive with a gain of unity. On negative input signals IC_2 gives a positive output at unity gain from the direct input, and zero output from IC_1, thus giving an aggregate output that is positive with unity gain. The circuit thus gives a unity-gain output to both positive and negative input half-cycles, and therefore acts as a full-wave rectifier. For correct operation of this circuit as a precision full-wave rectifier, it is important that resistors R_2 to R_6 be closely matched high stability types.

Figure 3.12 shows how the *Figure 3.11* circuit can be modified so that it acts as a precision full-wave a.c./d.c. converter which gives a positive d.c. output equal to the r.m.s. value of a sine wave input. The circuit is identical to that of *Figure 3.11*, except that the value of R_4 is increased to 11.1 kΩ to give form-factor correction, and that integrating capacitor C_1 is added between the output and the negative input of IC_2.

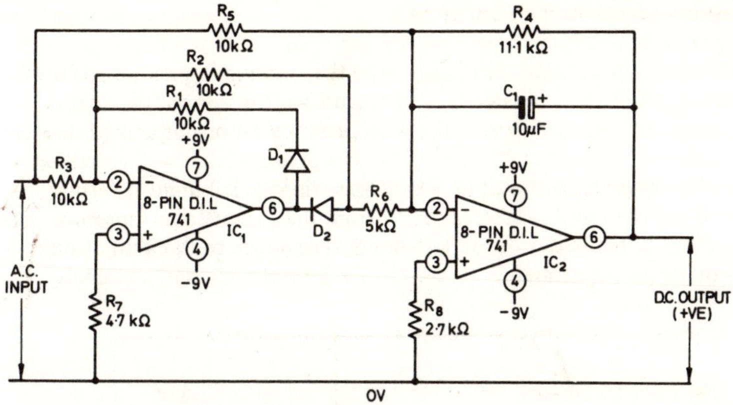

Figure 3.12. Precision full-wave a.c./d.c. converter.

The circuit can be used as a unity-gain converter for inputs up to 2 V r.m.s., over the frequency range 20 Hz to 15 kHz. The accuracy of the converter depends on the accuracies of the R_2 to R_6 resistors.

Finally, *Figure 3.13* shows an alternative way of using op-amps to make precision full-wave rectifiers. In this case IC_1 is used as a half-wave rectifier, but gives both positive and negative outputs, which are fed to IC_2. On positive input half-cycles IC_1 applies a negative output to the negative terminal of IC_2, and applies zero input to the positive terminal. Under this condition IC_2 acts as a unity-gain inverting amplifier, and gives a positive output. On negative input half-cycles IC_1 applies a

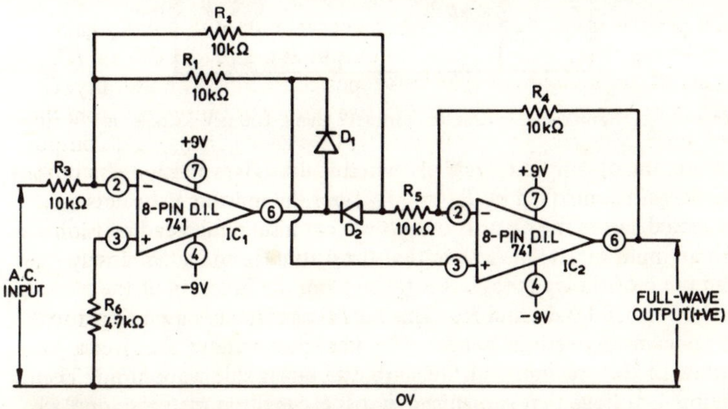

Figure 3.13. Alternative precision full-wave rectifier.

positive output to the positive terminal of IC_2, and applies zero input to the negative terminal. Under this condition IC_2 acts as a unity-gain non-inverting amplifier or voltage follower, and again gives a positive output. The circuit thus gives a unity-gain positive output to both positive and negative input half-cycles, and therefore acts as a full-wave rectifier.

The *Figure 3.13* circuit uses two resistors less than *Figure 3.11*, and has twice the input resistance. The *Figure 3.13* circuit can thus be regarded as the better of the two designs, although both circuits give identical performance.

D.C. volt and current meter circuits

Op-amps can readily be connected to moving-coil meters in such a way that they act as precision d.c. voltmeters, millivoltmeters, or current meters. If required, op-amps can be used as converters to extend the ranges of existing moving-coil d.c. voltmeters. *Figure 3.14* shows the practical circuit of a simple converter that enables a 1 V d.c. meter to read 100 mV full scale.

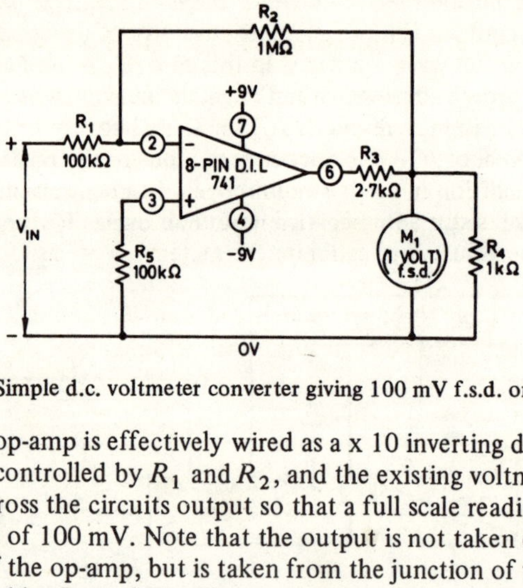

Figure 3.14. Simple d.c. voltmeter converter giving 100 mV f.s.d. on a 1 V meter.

Here, the op-amp is effectively wired as a x 10 inverting d.c. amplifier, with its gain controlled by R_1 and R_2, and the existing voltmeter is connected across the circuits output so that a full scale reading is obtained with an input of 100 mV. Note that the output is not taken directly from pin 6 of the op-amp, but is taken from the junction of the potential divider formed by R_3 and R_4. This potential divider is used to protect the meter from overload danger. The maximum voltage that can appear at pin 6 of the op-amp, with the amplifier saturated, is about 8 V, so the maximum voltage that can appear across R_4 and the meter is approximately 2 V. This voltage is too small to damage the meter, but great enough to ensure that full scale deflection can be linearly obtained under normal operating conditions. The voltmeter used in this circuit can be any 1 V type with a sensitivity greater than 1 kΩ/V.

Figure 3.15 shows an improved version of the d.c. voltmeter converter. The circuit is similar to that of *Figure 3.14*, except that the R_1 value can be selected to give full scale voltage values between 1 mV and 10 V (as shown in the table), and that offset-null control R_6 is wired into the circuit to act as a 'set zero' control. The voltmeter used in this circuit can again be any 1 V type with a sensitivity greater than 1 kΩ/V.

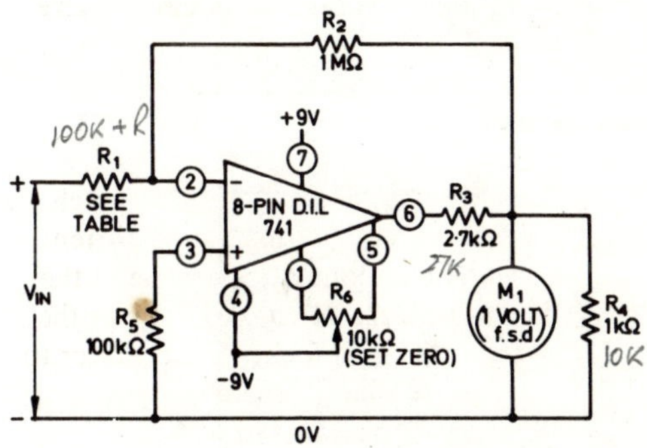

$V_{f.s.d}$	R_1
10V	10MΩ
1V	1MΩ
100MV	100kΩ
10MV	10kΩ
1MV	1kΩ

Figure 3.15. High-performance d.c. voltmeter converter.

Note in both of the above circuits that the converter has an effective sensitivity of 1 MΩ/V, irrespective of the sensitivity of the actual voltmeter used. Also note that the accuracy of the converter is dictated by the accuracies of R_1 and R_2. If required, R_2 can be replaced by a 820 kΩ fixed resistor and a 500 kΩ variable resistor in series: the variable resistor can be adjusted to calibrate the meter to a precise full scale voltage value.

Figure 3.16 shows how an op-amp and a moving-coil d.c. current meter can be wired together to form a precision d.c. volt or millivolt meter. The meter can be any type with a full-scale sensitivity in the range 100 μA to 5 mA, and the circuit can be made to give any decade value of full scale voltage reading in the range 1 mV to 1 000 V. The two tables show the component values that must be used to suit different meter types and to give alternative full scale voltage readings. The circuit has a basic sensitivity of 1 MΩ/V, irrespective of the meter type used.

Basically, the *Figure 3.16* circuit is wired as an inverting d.c. amplifier, with its output voltage appearing across R_4 and its gain determined by R_1 and R_2. The meter is wired in series between the $R_2 - R_4$ junction and the pin 6 output terminal of the op-amp, and gives a reading equal to the total output current of the op-amp. Since R_2 is large relative to R_4, the magnitude of the output current is dictated almost entirely by the magnitudes of R_4 and the output voltage, and is directly proportional to

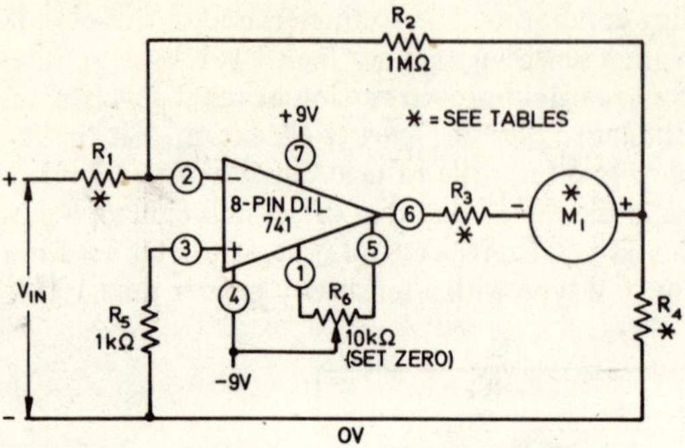

M_1	R_4	R_3
100µA	10kΩ	27kΩ
500µA	2kΩ	5·6kΩ
1MA	1kΩ	2·7kΩ
2·5MA	400Ω	1·5kΩ
5MA	200Ω	470Ω

$V_{f.s.d}$	R_1
1000 V	1000MΩ
100V	100MΩ
10V	10MΩ
1V	1MΩ
100MV	100kΩ
10MV	10kΩ
1MV	1kΩ

Figure 3.16. D.C. volt/millivolt meter, with alternative component values.

the input voltage. Offset-null control R_6 is used as a *set zero* control. If required, R_2 can be replaced by an 820 kΩ fixed resistor and a 500 kΩ variable resistor in series, to facilitate precise calibration of the meter.

Figure 3.17 shows an alternative way of using an op-amp and a moving-coil meter as a simple voltmeter or as a current meter. In this case the op-amp is wired as a unity-gain non-inverting d.c. amplifier or voltage follower, with its output appearing across R_4. The circuit uses a 1 mA f.s.d. meter as a readout indicator, and this meter reads the value of the current flowing into R_4 from the op-amp output. Consequently, the meter reading is directly proportional to the magnitude of the input voltage appearing across R_2, and equals full scale with an input of 100 mV.

Potential divider $R_1 - R_2$ enables the input voltage to be ranged so that the circuit acts as a voltmeter giving full scale decade readings from 100 mV to 1 000 V, or as a current meter giving full scale decade readings from 1 µA to 1 A. The two tables show suitable R_1 and R_2 values for using the circuit as a volt or current meter. If required, R_4 can be relaced by an 82 Ω fixed resistor and a 50 Ω variable resistor in series, to facilitate precise calibration of the meter.

Figure 3.18 shows how the *Figure 3.17* circuit can be modified so that it acts as a precision multi-range d.c. millivoltmeter, with a maximum sensitivity of 1 mV full scale. In this case the op-amp is wired as a non-inverting x 1 000 d.c. amplifier, with gain controlled by R_8 and R_9, and

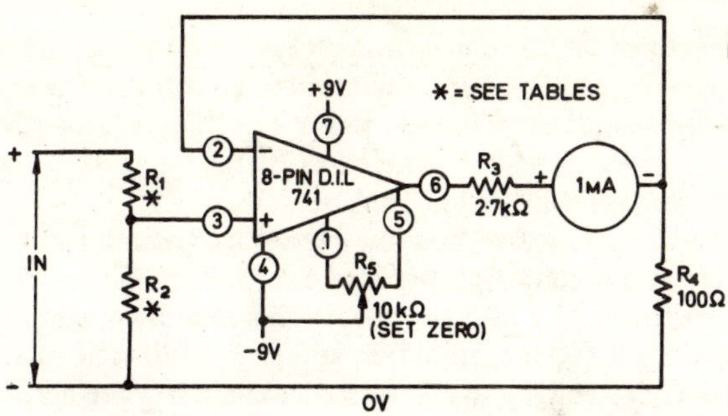

R3 100uA 27K
500 5·6K
1MA 2·7K
2·5 1·5K
5 470Ω

R4 100uA 10K
500 2K
1 MA 1K
2·5 400Ω
5 200Ω

VOLTMETER		
f.s.d	R_1	R_2
1000V	10MΩ	1kΩ
100V	10MΩ	10kΩ
10V	10MΩ	100kΩ
1V	900kΩ	100kΩ
100mV	0Ω	100kΩ

CURRENT METER		
f.s.d	R_1	R_2
1A	0Ω	0·1Ω
100mA	0Ω	1Ω
10mA	0Ω	10Ω
1mA	0Ω	100Ω
100µA	0Ω	1kΩ
10µA	0Ω	10kΩ
1µA	0Ω	100kΩ

Figure 3.17. Simple d.c. voltage or current meter.

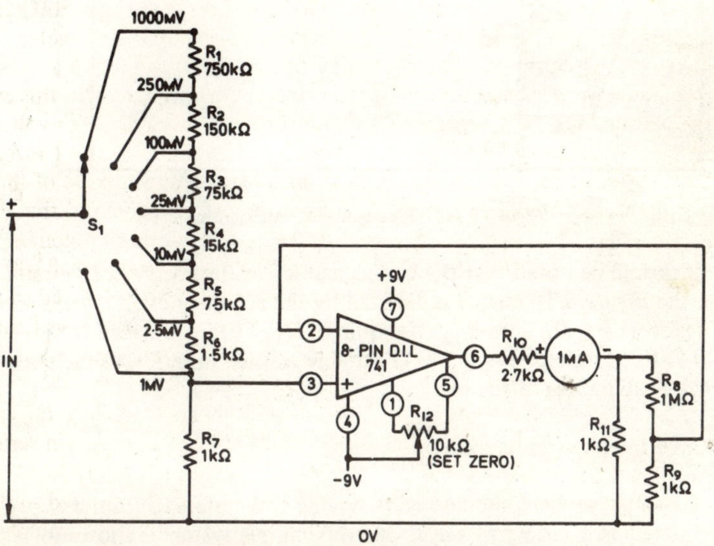

Figure 3.18. Precision d.c. millivoltmeter.

the meter measures the current flowing in load resistor R_{11}. The meter is a 1 mA f.s.d. type, and R_{11} has a value of 1 kΩ, so the full scale indication corresponds to an output of 1 V or an input (across R_7) of 1 mV. The R_1 to R_7 potential divider enables the input to be ranged to give full scale voltage values from 1 mV to 1 V.

Finally, *Figure 3.19* shows how the *Figure 3.18* circuit can be modified so that it acts as a precision multi-range d.c. microammeter, with a maximum sensitivity of 1 μA full scale. In this case the op-amp is wired as a non-inverting x 100 d.c. amplifier, and gives a full scale reading with an input of 10 mV. The R_1 to R_7 resistance chain is wired across the input of the op-amp as a tapped current shunt, and developes an input voltage proportional to the magnitude of the input current. Thus, the meter reading is directly proportional to the magnitude of the input current.

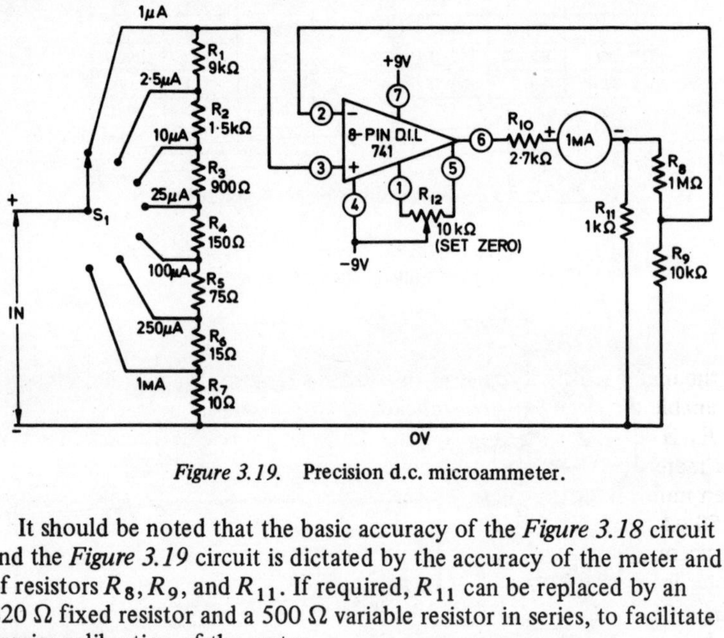

Figure 3.19. Precision d.c. microammeter.

It should be noted that the basic accuracy of the *Figure 3.18* circuit and the *Figure 3.19* circuit is dictated by the accuracy of the meter and of resistors R_8, R_9, and R_{11}. If required, R_{11} can be replaced by an 820 Ω fixed resistor and a 500 Ω variable resistor in series, to facilitate precise calibration of the meter.

A.C. voltmeter circuits

Op-amps can be combined with moving-coil meters to form a.c. voltmeters in a variety of ways. One very simple system is shown in *Figure 3.20*. In this case the op-amp is wired as an inverting d.c. amplifier, with R_1 acting as an input resistor, and the meter and the D_1 to D_4

bridge rectifier acting as the negative feedback element. In this configuration the currents in R_1 and the feedback element are always equal. A virtual earth exists between ground and the negative input terminal of the op-amp, so the signal current of R_1 (and thus of the meter) is directly proportional to the magnitude of the input signal voltage, and the meter gives a perfectly linear reading of the input voltage.

In effect, the op-amp in the *Figure 3.20* circuit is used purely to linearise the characteristics of the bridge rectifier, and gives no increase

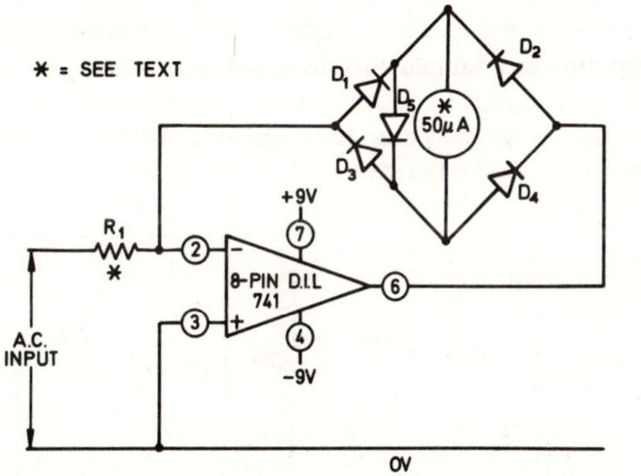

NOTE: D_1 TO D_4 = GENERAL-PURPOSE GERMANIUM DIODES
D_5 = GENERAL-PURPOSE SILICON DIODE

Figure 3.20. Simple a.c. voltmeter.

in the meter sensitivity, and R_1 is used as a ranging or multiplier resistor to enable the circuit to give indications in terms of a.c. voltage. The value of R_1 is determined by multiplying the basic ohms/volt d.c. sensitivity of the meter by 0.9 to establish an a.c. ohms/volt sensitivity figure, and by then multiplying this figure by the required full scale voltage value, e.g. A 50 μA meter has a d.c. sensitivity of 20 kΩ/V, so the a.c. sensitivity works out at 18 kΩ/V. R_1 must thus be given a value of 18 kΩ if the meter is required to read 1 V f.s.d., or 1.8 MΩ if the meter is required to read 1 000 V f.s.d.

In practice, the *Figure 3.20* circuit can be used with any moving coil meter giving a full scale value in the range 50 μA to 1 mA, and R_1 can be selected to give any required full scale voltage value in the approximate range 100 mV to 1 000 V. The voltmeter gives a useful performance at frequencies up to about 40 kHz. If the circuit is to be used to measure a.c. voltages imposed on d.c., a capacitor must be wired in series with the input terminals to block the d.c. component.

The input impedance of the *Figure 3.20* circuit is equal to the R_1

value, and may be quite low. In many applications it may be essential to use a voltmeter having a high input impedance, and in such cases a circuit of the type shown in *Figure 3.21* can be used. This circuit has a typical input impedance of tens or hundreds of Megohms, and gives a useful performance up to frequencies of about 40 kHz.

The operating theory of the *Figure 3.21* circuit is quite simple. The op-amp is wired as a unity-gain a.c. voltage follower, with its output appearing across load resistor R_1, and with input resistor R_2 bootstrapped so that it appears as a virtual open circuit to a.c. signals. The meter and bridge rectifier are effectively wired in series with the output of the op-amp, and the meter reads the a.c. signal current flowing in load resistor R_1. The magnitude of this current is directly proportional to the magnitude of the a.c. input signal voltage, so the circuit acts as a highly linear a.c. voltmeter, and has a very high input impedance.

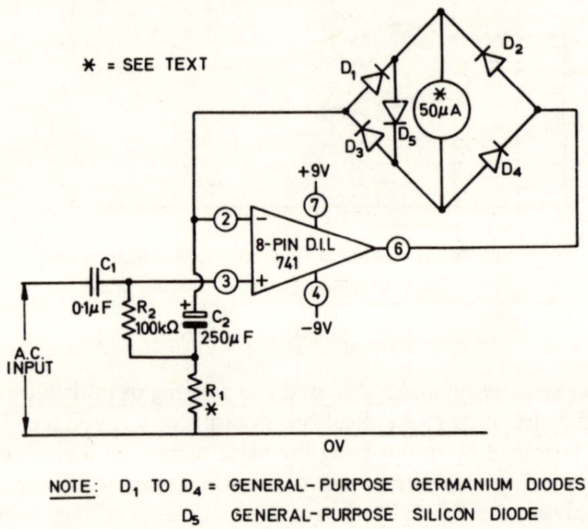

Figure 3.21 High-impdance a.c. voltmeter.

The *Figure 3.21* circuit can be used with any moving-coil meter giving a full scale value in the range 50 μA to 1 mA, and R_1 can be selected to give any full scale voltage value in the range 100 mV to 1 V. If higher voltage ranges are required, they can be obtained by connecting the test voltage to the input of the op-amp via a ranging potential divider. The procedure for selecting the R_1 value is the same as that described for the *Figure 3.20* circuit. Thus, if the circuit is required to read 1 V full scale when used with a 50 μA meter, R_1 must be given a value of 18 kΩ, and if the circuit is required to read 1 V full scale when used with a 1 mA meter, R_1 must be given a value of 900 Ω. Using a 50 μA meter, the prototype circuit gives an input impedance of 33 MΩ in parallel with 18 pF when R_1 is given a value of 1.8 kΩ for 100 mV full scale, and gives an input impedance of 330 MΩ in parallel with 5 pF when R_1 is given a value of 18 kΩ for 1 V full scale.

It should be noted in the *Figure 3.20* and *3.21* circuits that diode D_5 is wired directly across the indicating meter within the bridge rectifier. This diode is used to protect the meter against overload damage, and can be any general-purpose silicon type.

Finally, *Figure 3.22* shows the circuit of a precision a.c. volt or millivolt meter. The circuit can be used with any moving coil meter giving a full scale reading in the range 100 μA to 5 mA, and can be made to give any full scale a.c. voltage reading in the range 1 mV to 1 000 V.

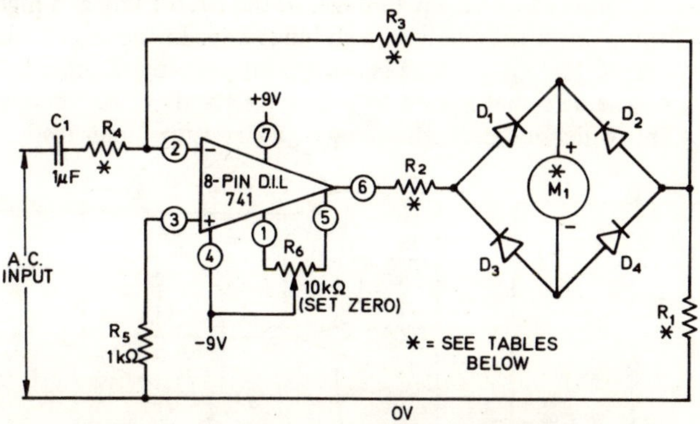

NOTE: D1 TO D4 = GENERAL-PURPOSE GERMANIUM DIODES

M1	R1	R2
100μA	9kΩ	27kΩ
500μA	1·8kΩ	5·6kΩ
1MA	900Ω	2·7kΩ
2·5MA	360Ω	1·5kΩ
5MA	180Ω	470Ω

VALUES FOR USE WITH DIFFERENT METER MOVEMENTS

$V_{f.s.d}$	R4	R3
1000V	10MΩ	10 kΩ
100 V	10MΩ	100 kΩ
10V	10MΩ	1MΩ
1V	1MΩ	1MΩ
100MV	100kΩ	1MΩ
10MV	10kΩ	1MΩ
1MV	1kΩ	1MΩ

DIFFERENT f.s.d VOLTAGE SENSITIVTIES

Figure 3.22 Precision a.c. volt/millivolt meter.

The operating theory of the circuit is quite simple; the op-amp is wired as an inverting amplifier in which the meter reads the magnitude of the a.c. current flowing in load resistor R_1. The magnitude of this current is directly proportional to the input voltage of the circuit and to the gain of the circuit, so the circuit acts as a linear a.c. volt or millivolt meter. The circuit is in fact designed so that the meter reads full scale when 1 V appears across R_1, so the full scale voltage reading is dictated by the ratios of R_3 and R_4. One of the tables in *Figure 3.22* shows suitable R_3 and R_4 values for alternative full scale decade voltages from 1 mV to 1 000 V.

The circuit has automatic overload protection provided by limiting resistor R_2, and the second table in *Figure 3.22* shows suitable R_1 and R_2 values for use with alternative meter types. Offset-null control R_6 is used in the circuit as a set-zero control.

Linear-scale ohmmeter circuits

Figure 3.23 shows how an op-amp can be connected to a moving-coil meter to form a linear-scale ohmmeter giving full scale readings in the range 1 kΩ to 10 MΩ. The circuit is divided into two parts, and consists of a voltage generator that is used to generate a standard test voltage, and a readout unit which indicates the value of the resistor under test.

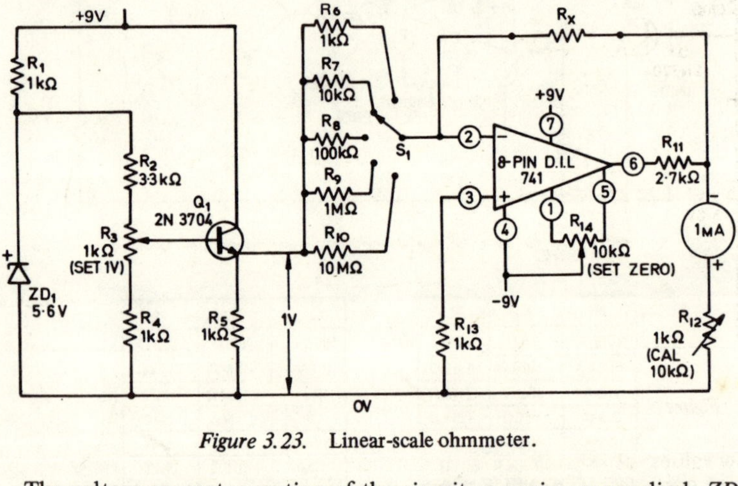

Figure 3.23. Linear-scale ohmmeter.

The voltage generator section of the circuit comprises zener diode ZD_1, transistor Q_1, and resistors R_1 to R_5. The action of these components is such that a stable reference potential of 1 V is developed across R_5, but is adjustable over a limited range via R_3. This reference voltage is fed to the input of the op-amp readout unit. The op-amp is wired as an inverting d.c. amplifier, with the 1 mA meter and R_{12} forming a 1 V f.s.d. meter across its output, and with the op-amp gain determined by the values of ranging resistors R_6 to R_{10} and by negative feedback resistor R_x. Since the input to the amplifier is fixed at 1 V, the output voltage reading of the meter is directly proportional to the value of R_x, and equals full scale when R_x and the ranging resistor values are equal. Consequently, the circuit functions as a linear-scale ohmmeter.

The circuit has five full scale decade ranges, from 1 kΩ to 10 MΩ. The linearity of the meter is excellent on all but the 10 MΩ range, where linearity errors may be as great as 10 %. The full-scale accuracy of the circuit is dependent on the accuracies of ranging resistors R_6 to R_{10}.

The procedure for initially calibrating the *Figure 3.23* circuit is as follows. First, switch the unit to the 10 kΩ range and secure an accurate 10 kΩ resistor in the R_x position. Now adjust R_3 to give an accurate 1 V across R_5, and then adjust R_{12} to give a precise full scale reading on the meter. All adjustments are then complete, and the circuit is ready for use.

Figure 3.24 shows an alternative type of linear-scale ohmmeter circuit. This particular design is specifically intended to give accurate readings of

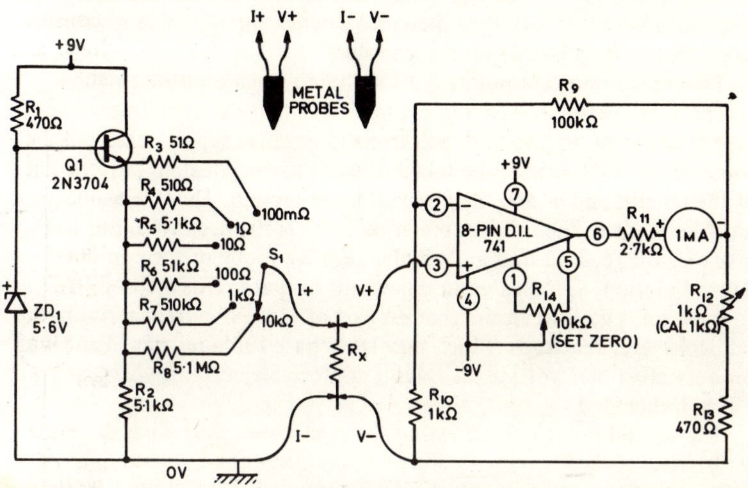

Figure 3.24. Low-value linear ohmmeter, with probe details shown in inset.

low values of resistance, and its most sensitive range reads 100 MΩ full-scale. The design is such that the accuracy of the reading is not influenced by the resistance of the actual connecting leads that are used to connect the meter to the resistor under test.

The circuit is made up of two separate sections. One of these is a constant current generator, which is used to pass a fixed current through the resistor under test, and the other is a d.c. millivoltmeter, which measures the voltage that is developed across the test resistor by the fixed current. Since the test current is fixed, the magnitude of this voltage is directly proportional to the resistance of R_x, and the circuit thus functions as a linear-scale ohmmeter.

The d.c. millivoltmeter or readout section of the circuit is designed around the op-amp, which functions as a non-inverting x 100 d.c. amplifier. The circuit requires an input of 10 mV for full-scale deflection on the meter, and the input voltage is applied to the high-impedance positive terminal of the op-amp.

The constant current section of the unit is designed around emitter-follower transistor Q_1, which has a zener-stabilised potential of 5.6 V applied to its base and thus generates a stable reference potential of approximately 5 V at its emitter. This reference potential is fed to R_x via ranging resistors R_3 to R_8, which determine the current flowing in R_x. Since only 10 mV needs be developed across R_x to give a full scale reading on the meter, the R_x voltage has negligible effect on the voltage developed across the ranging resistors or the current flowing in them, and these resistors thus effectively act as constant current test sources. These test currents in fact vary by a mere 0.2 % when the R_x value is varied from zero to its full-scale resistance value.

Two very important points must be noted when constructing the *Figure 3.24* circuit. The first point to note is that the circuit uses two supply batteries, to give both positive and negative supply rails, and the common or 0 V line *must* be taken directly to the constant current side of the circuit, and *not* to the millivoltmeter section. The 0 V connection must be made to the millivoltmeter section via the negative input test probe of the readout unit and via the negative probe of the constant current section, as indicated in the circuit diagram. This method of connection is used to ensure that no part of the test current flows in the millivoltmeter test leads, which inevitably have finite resistance and will produce their own voltage readings if appreciable currents are passed through them.

The second point to note is that the circuit uses *four* separate test leads. Two of these leads (*I+* and *I–*) came from the constant current section, and are used to apply the test current to R_x, and the other two (*V+* and *V–*) go to the readout unit, and are used to measure the voltage directly across R_x. In practice, these four leads can be terminated in a pair of brass or similarly conductive test prods, as shown in the inset of *Figure 3.24*. It is important that the readout tests probes (*V+* and *V–*) be secured as close as possible to the tips of the test probes.

Once construction of the circuit is complete, the unit can be tested and calibrated. To test the unit, first switch it to the 10 Ω range, short the test probes together, and check that the meter can be set to zero via R_{14}. If all is well, remove the short from the probes and check that the meter reading goes beyond full scale. Now reconnect the short to the probes, turn to the 100 mΩ range, and check that a minimum reading of less than 2 mΩ can be obtained on the meter, adjusting R_{14} as necessary. If an appreciably greater reading than 2 mΩ is obtained, it is probable that the 0 V line has been incorrectly connected to the circuit, or that the construction of the test probes is faulty.

When all is well, the unit can be calibrated by switching it to the 1 kΩ range, connecting a 1 kΩ resistor in the R_X position, and then adjusting R_{12} to obtain full scale deflection on the meter. All adjustments are then complete, and the unit is ready for use, since the accuracy of the meter on all remaining ranges is determined primarily by the accuracies of ranging resistors R_3 to R_8.

A linear-scale capacitance meter

To conclude this chapter, *Figure 3.25* shows the basic circuit of a linear-scale capacitance meter which can give full-scale readings in the range 100 pF to 1 μF when powered from a suitable sine wave source. The operating principle of the circuit is quite simple. The op-amp is wired as an inverting a.c. amplifier, with the meter and bridge rectifier connected to its input in such a way that the meter reads full scale when approximately 1 V r.m.s. appears at the $R_1 - R_2$ output junction. The gain of the op-amp is determined by the relative impedances of R_1 and C_X, and equals unity when these two impedances are equal.

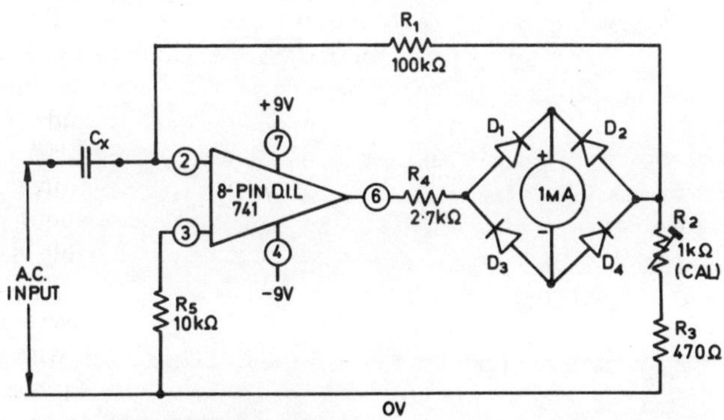

f_{IN}	$V_{IN(RMS)}$	$C_{f.s.d}$
≏ 18 kHz	≏ 1V	100pF
≏ 1·8kHz	≏ 1V	1000pF
≏ 180 Hz	≏ 1V	·01μF
≏ 18Hz	≏ 1V	0·1μF
≏ 18Hz	≏ 100mV	1μF

Note: D_1 to D_4 = GENERAL PURPOSE GERMANIUM DIODES

Figure 3.25. Basic circuit of linear-scale capacitance meter, with details of approximate input signal requirements for full-scale decade capacitance ranges from 100 pF to 1 μF.

Suppose then that a 1 V r.m.s. sine wave signal is applied to the input of the circuit, and that this signal is adjusted to a frequency (approximately 1.8 kHz) at which the impedance of a 1 000 pF capacitor equals

100 kΩ. If a 1 000 pF capacitor is now placed in the C_x position the meter will read full scale, since the op-amp gives unity-gain under this condition. If, on the other hand, the C_x value is reduced by a factor of ten, to 100 pF, its impedance will rise by a factor of ten, to 1 MΩ, and the circuit gain will fall to 0.1, and the meter will read only 1/10th of full scale. The meter reading is thus directly proportional to the C_x capacitor value, and the circuit functions as a linear-scale capacitance meter.

To make the meter read alternative full-scale capacitance values, it is simply necessary to adjust the input frequency and/or amplitude to the approximate values shown in the table of *Figure 3.25*, and to then place an accurate capacitor with the required f.s.d. value in the C_x position. Adjust R_2 to give a full scale indication on the meter, remove the calibration capacitor, and the circuit is ready for use as a linear-scale capacitance meter. The meter can be used in conjunction with any reasonable sine wave generator having an output impedance less than 1 kΩ or so.

CHAPTER 4

20 OSCILLATOR AND MULTIVIBRATOR PROJECTS

Operational amplifiers can be used in a variety of waveform generating applications. They can readily be made to function as oscillators and multivibrators, and can be used to generate sine waves, square waves, triangle waves, and waves with a variety of ramp and pulse forms. They can readily be used to generate waveforms at repetition rates as low as a few cycles per hour or as high as 20 000 cycles per second.

Twenty useful op-amp oscillator and multivibrator projects of various types are described in the present chapter. All of these projects are designed around the popular type 741 integrated-circuit op-amp, and the pin connections shown in the following diagrams apply to the 8-pin dual-in-line version of the device only.

Sine-wave oscillations

Low frequency sine waves can be generated in a variety of ways. One very simple method is to wire a critically adjusted twin-T network between the output and the input of an inverting operational amplifier, as shown in *Figure 4.1*.

Here, the twin -T network comprises $R_1 - R_2 - R_3 - R_4$ and $C_1 - C_3 - C_4$. In a normal twin-T circuit the network is said to be balanced when its components are in the ratios $R_1 = R_2 = 2(R_3 + R_4)$, and $C_1 = C_2 = C_3/2$. When the network is perfectly balanced it acts as a frequency-dependent attenuator, and gives zero output at a centre frequency equal to $1/6.28 R_1 C_1$, and a finite output at all other frequencies. When the twin-T network is imperfectly balanced, it gives an attenuated but finite output at the centre frequency, and the phase of this output signal depends on the direction of the imbalance: if the imbalance is caused by $(R_3 + R_4)$ being too low in value, the phase of the output is inverted relative to the input.

In the *Figure 4.1* circuit the input of the twin-T network is taken from the output of the op-amp, the output of the twin-T is fed to the inverting input terminal of the op-amp, and R_4 is critically adjusted so that the twin-T gives a small output at the centre-frequency, this output being phase-inverted relative to the input. Thus, zero overall phase inversion takes place between the output and the input of the op-amp at the centre-frequency, and the circuit oscillates at the centre-frequency of the twin-T network. With the component values shown, the circuit oscillates at a frequency of approximately 1 kHz.

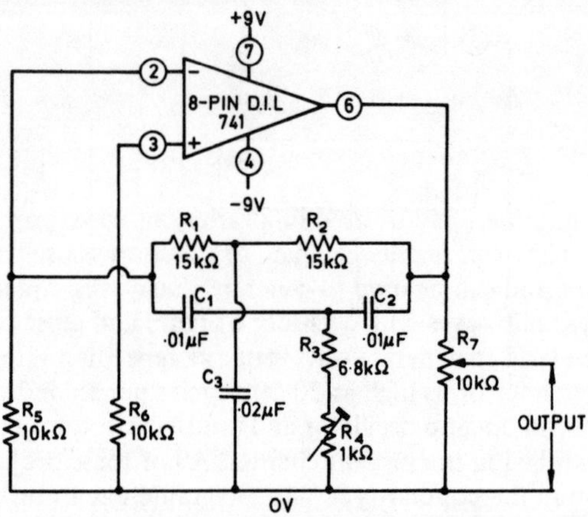

Figure 4.1. 1 kHz twin-T test oscillator.

The output amplitude of the circuit is fully variable from zero to approximately 5 V r.m.s. via R_7. In use, R_4 should be adjusted so that the circuit only just oscillates, under which condition the output typically contains less than 1 % total harmonic distortion. Automatic amplitude control is obtained in this circuit because of the progressive non-linearity of the op-amp as the output signal approaches the clipping level.

An alternative method of automatic amplitude control is shown in the 1 kHz oscillator circuit of *Figure 4.2*. In this case silicon diode D_1 is wired between the output and the input of the op-amp via potential divider R_7. The diode progressively conducts and reduces the voltage gain of the amplifier circuit when the voltage across the diode exceeds a few hundred millivolts, and thus functions as an automatic amplitude control.

To set up the *Figure 4.2* circuit, first set R_7 so that its slider is at the op-amp output end of the pot. Now adjust R_4 so that oscillation ceases, and then slowly advance R_4 just past the point at which oscillation begins again. Under this condition the sine wave output signal has an amplitude of approximately 500 mV peak-to-peak, or 170 mV r.m.s., and all adjustments are complete. R_7 then enables the output signal, which contains negligible distortion, to be varied between 170 mV and 3 V r.m.s.

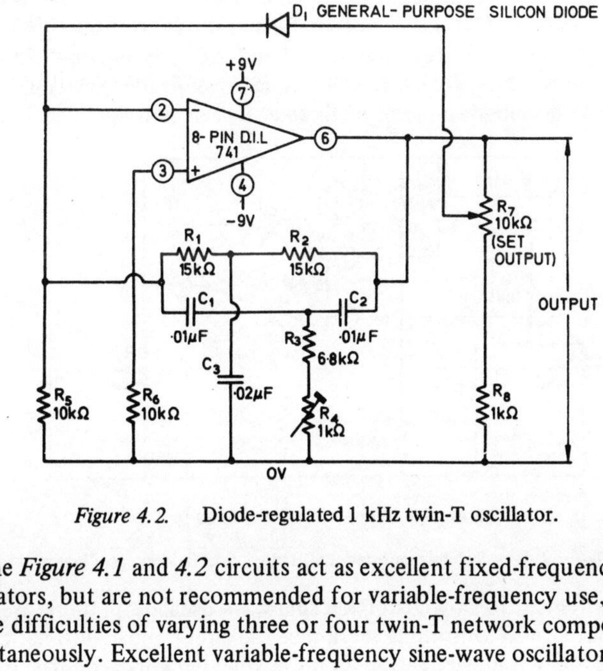

Figure 4.2. Diode-regulated 1 kHz twin-T oscillator.

The *Figure 4.1* and *4.2* circuits act as excellent fixed-frequency oscillators, but are not recommended for variable-frequency use, because of the difficulties of varying three or four twin-T network components simultaneously. Excellent variable-frequency sine-wave oscillators can,

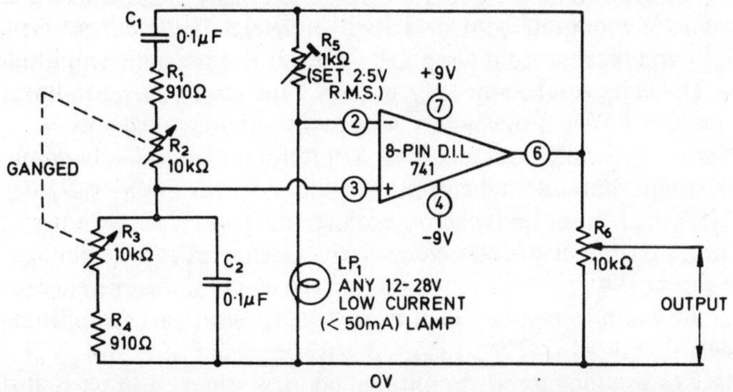

Figure 4.3. 150 Hz - 1.5 kHz Wien-bridge oscillator.

however, be made by using Wien frequency-selective networks in conjunction with op-amps, as shown in the circuits of *Figures 4.3* to *4.5*.

The operating frequencies of these three circuits can be varied over a decade range via twin-gang variable resistors R_2 and R_3: the circuits differ only in the methods used for obtaining automatic amplitude control. In all cases the Wien network is connected between the output and the non-inverting input terminal of the op-amp, and comprises $R_1 - R_2 - R_3 - R_4$ and $C_1 - C_2$, and an automatic amplitude-controlling potential divider is connected between the output and the inverting input of the op-amp. The Wien network acts as a frequency-dependent attenuator that gives an attenuation factor of three at its centre frequency. The

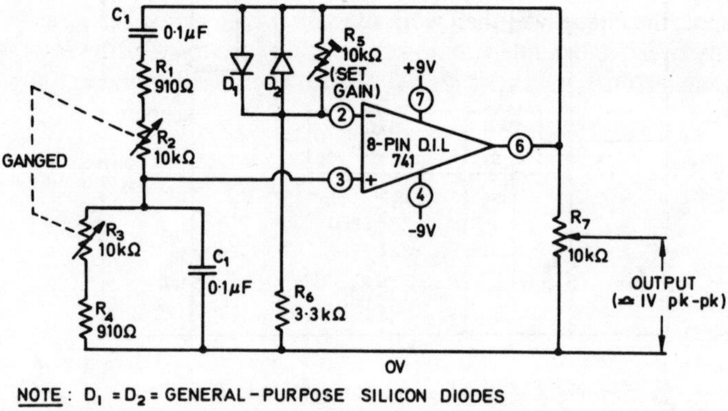

Figure 4.4. Diode-regulated 150 Hz - 1.5 kHz Wien oscillator.

basic requirement for low-distortion sine wave oscillation, therefore, is that the amplitude-controlling section of the circuit must automatically give fractionally less attenuation than the Wien network, and thus ensure that the overall gain of the circuit is only fractionally greater than unity.

Automatic amplitude control is obtained in the *Figure 4.3* circuit by wiring R_5 and incandescent lamp LP_1 in series as a self-adjusting potential divider. The lamp can be any 12 V to 28 V type with a current rating of less than 50 mA. When the circuit is correctly set up, the sine wave output signal typically contains about 0.1 % total harmonic distortion, and the circuit draws a total current of about 6 mA from the supply lines. The circuit is set up by simply adjusting R_5 to give approximately 2.5 V r.m.s. output at the maximum output setting of R_6.

The *Figure 4.4* and *4.5* circuits rely on the onset of diode or zener diode conduction to regulate their overall gains. Both circuits inevitably produce a slightly (1 or 2 %) distorted waveform, but offer the great advantage of producing zero amplitude 'bounce' when range sweeping in variable-frequency circuits. The maximum peak-to-peak output of each circuit is roughly double the breakdown voltage of the semiconductor regulation element. In the *Figure 4.4* circuit the diodes begin to conduct at 500 mV, so this circuit gives a maximum peak-to-peak output of only 1 V. In the *Figure 4.5* circuit, on the other hand, zener diodes ZD_1 and ZD_2 are connected back-to-back and may have values as high as 5.6 V, so the maximum peak-to-peak output may be as high as 12 V or so.

The procedure for setting up the *Figure 4.4* and *4.5* circuits is as follows. First, adjust the SET GAIN control so that the circuit just goes into stable operation, with minimum distortion. Now sweep the frequency band and check that oscillation is obtained over the whole range of the circuit. If necessary, find any weak spots in the band and then adjust the SET GAIN control so that good oscillation is obtained at that spot; the circuit will then work over the whole band. The level stability over the band depends on the tracking accuracy of the $R_2 - R_3$ two-gang resistors, and these should be high quality components for best results.

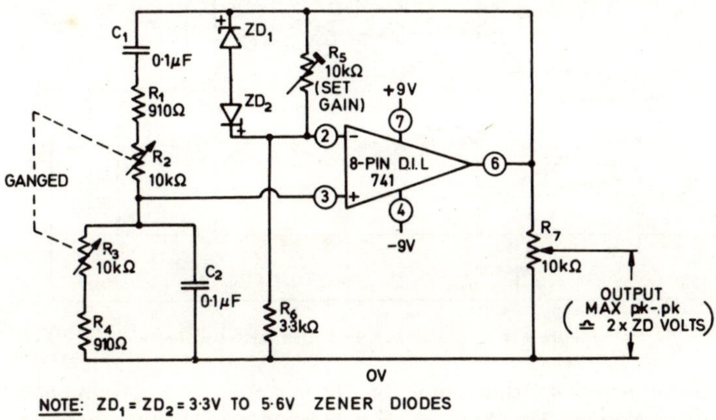

Figure 4.5. Zener-regulated 150 Hz - 1.5 kHz Wien oscillator.

The *Figure 4.3* to *4.5* circuits are designed to cover the range 150 Hz to 1.5 kHz with the component values shown. If required, the frequency range can be changed by using different C_1 and C_2 values; increasing the capacitor values lowers the frequency. The maximum available operating frequency (for a low distortion output) of each circuit is limited to about 25 kHz, due to the slew-rate limitations of the 741 operational amplifier.

The Wien oscillator circuits that we have looked at can be modified in a number of ways to meet specific performance requirements. They can, for example, be adapted for use as fixed-frequency oscillators, or as adjustable fixed-frequency oscillators, or they can be simply adapted to operate from single (rather than dual) power supplies.

Figure 4.6, for example, shows how the *Figure 4.3* circuit can be modified to act as a 1 kHz single supply oscillator. R_7 and R_8 act as a potential divider, to give a quiescent half-supply rail voltage, and C_3 bypasses R_8 to a.c. and thus gives an effective low-impedance supply path for the circuit. Normally, with R_3 and R_4 out of the circuit, oscillation occurs at slightly less than 1 kHz. R_3 and R_4 are used to shunt the R_2 arm of the Wien network, and thus raise the operating frequency to precisely 1 kHz. If necessary, R_3 can be increased or decreased to bring the frequency to precisely 1 kHz if the tuning capacitors are substantially out of tolerance.

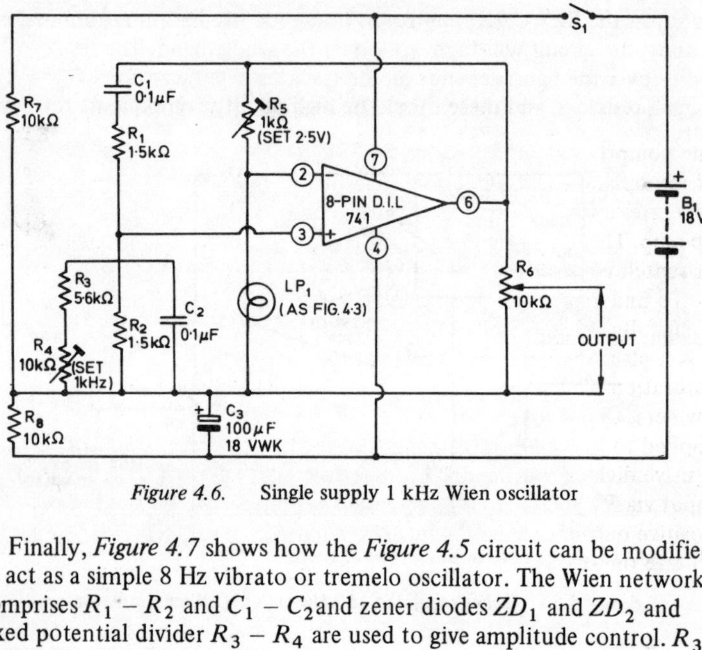

Figure 4.6. **Single supply 1 kHz Wien oscillator**

Finally, *Figure 4.7* shows how the *Figure 4.5* circuit can be modified to act as a simple 8 Hz vibrato or tremelo oscillator. The Wien network comprises $R_1 - R_2$ and $C_1 - C_2$and zener diodes ZD_1 and ZD_2 and fixed potential divider $R_3 - R_4$ are used to give amplitude control. R_3 has a value slightly greater than double that of R_4, to ensure reliable oscillation with little distortion.

Square-wave generators

Operational amplifiers can be made to act as excellent low-frequency square-wave generators by wiring them in the relaxation oscillator configuration shown in *Figure 4.8a.* Examination of this circuit shows it to contain two potential dividers, each driven from the output of the op-amp, and each with its output going to one or other of the op-amp input terminals. One of these potential dividers is of the resistive type

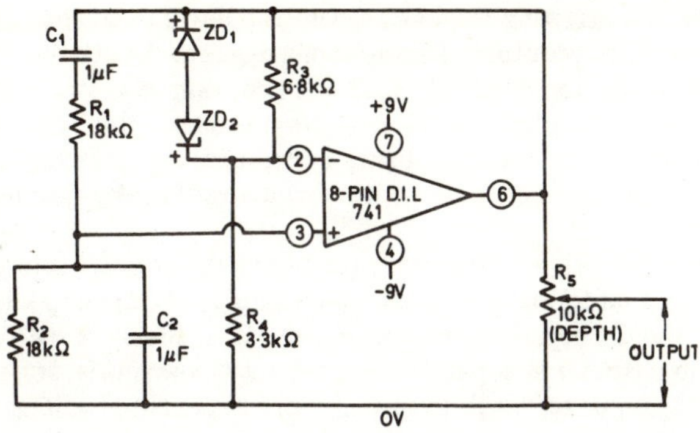

NOTE: ZD1 = ZD2 = 3·3V TO 5·6V ZENER DIODES

Figure 4.7. 8 Hz vibrato or tremelo oscillator.

and comprises R_2 and R_3, and feeds its output to the non-inverting input of the op-amp. The other potential divider comprises R_1 and C_1, and generates a timing waveform which is fed to the inverting input of the op-amp. The actual op-amp is used as a regenerative voltage comparator or switch which is activated by the relative levels of the two input signals.

To understand the operation of the basic circuit, assume initially that C_1 is fully discharged and that a regenerative switching action has just taken place in which the op-amp output has switched to a positive saturation level, thus applying a large positive voltage to both potential dividers. Under this condition half of the positive saturation voltage is applied to the non-inverting input of the op-amp via the $R_2 - R_3$ resitive divider, and a rising positive voltage is applied to the inverting input via R_1 and C_1 as C_1 charges up exponentially via R_1 and the positive output of the op-amp. As time passes the rising exponential voltage on the inverting terminal approaches, and eventually exceeds,

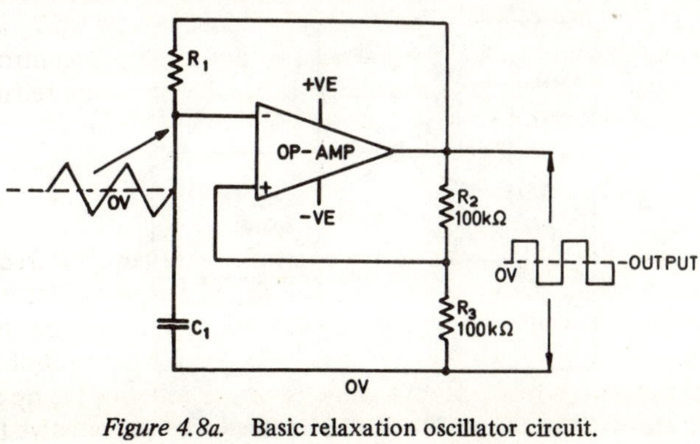

Figure 4.8a. Basic relaxation oscillator circuit.

that on the non-inverting terminal, at which point the op-amp comes out of saturation and its output starts to swing negative. Under this condition the voltage at the non-inverting input of the op-amp also swings negative via R_2 and R_3, but that at the inverting input tends to be held steady by the charge on C_1, the net result being that a sharp regenerative action takes place in which the op-amp output switches abruptly into negative saturation.

At the end of this regenerative stage half of the negative saturation voltage is applied to the non-inverting terminal of the op-amp via $R_2 - R_3$, and C_1 begins to recharge in a negative direction via R_1 and the op-amp output and applies a rising negative exponential voltage to the inverting input terminal. Eventually, this exponential voltage becomes slightly more negative than the voltage on the non-inverting terminal, at which point another regenerative action takes place and the op-amp output switches back to a positive saturation level. The whole timing and switching sequence then repeats ad infinitum. Thus, a series of square waves are generated at the output of the circuit, and a series of approximately triangular waves are developed across C_1.

This basic relaxation oscillator circuit has a number of interesting characteristics. The period of each half-cycle, and thus the operating frequency of the waveform, depends both on the time constant of the $C_1 - R_1$ network and on the voltage dividing ratios of R_2 and R_3, so the operating frequency can be changed by altering any one of these four component values. A wide-range variable-frequency square wave generator can thus be made by making just one of these components variable. The operating frequency of the circuit is dictated almost entirely by the values of the $R_1 - R_2 - R_3 - C_1$ components, and is virtually unaffected by variations in supply rail voltage, so the circuit has execellent frequency stability.

Figure 4.8b shows how the *Figure 4.8a* circuit can be modified to act

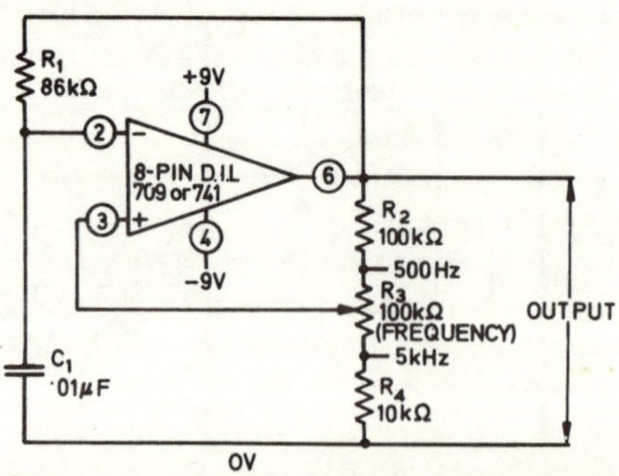

Figure 4.8b. Simple 500 Hz - 5 kHz square wave generator.

as a simple but effective variable-frequency square wave generator covering the approximate range 500 Hz to 5 kHz. In this case the frequency is made variable by adjustment of the division ratio of the $R_2 - R_3 - R_4$ potential divider. The division ratio, and thus the operating frequency, can be varied over a range of one decade. If required, the minimum operating frequency of the circuit can be set to precisely 500 Hz by adjusting the R_1 value, or by replacing R_1 with a 47 kΩ fixed resistor and a 100 kΩ variable resistor in series, as shown in the improved circuit of *Figure 4.9*. This circuit also shows how a 10 kΩ variable pot can be wired across the op-amp output to act as an output amplitude control.

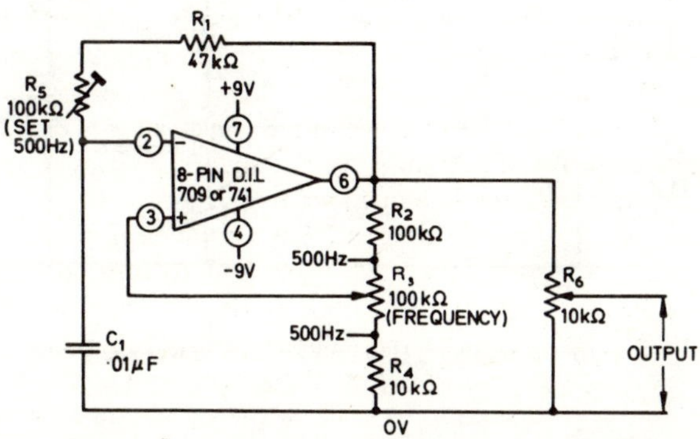

Figure 4.9. Improved 500 Hz - 5 kHz square wave generator.

Figure 4.10 shows how the *Figure 4.9* circuit can be further developed to cover the range 2 Hz to 20 kHz in four switched decades. Ranging is achieved by switch-selecting suitable timing capacitors and resistors. A pre-set variable resistor in each timing arm enables each range to be adjusted to give a precise frequency output at the minimum setting of the frequency control pot, thus giving accurate tracking over all four ranges of the frequency dial.

Finally, *Figure 4.11* shows how the basic square-wave generator can be adapted for use as a push-button tone generator. With the component values shown the circuit generates frequencies of 500 Hz via S_1, 670 Hz via S_2, and 760 Hz via S_3; alternatively frequencies can be obtained by changing the timing resistor values. This particular circuit can readily be used as the basis of a simple tone-signalling system, such as is used for remote-control purposes.

Note in the circuits of *Figure 4.8* to *Figure 4.11* that the op-amps used can be of either the 709 or 741 types. The 709 op-amp has a higher slew rate than the 741, and thus gives the best square-wave output. The

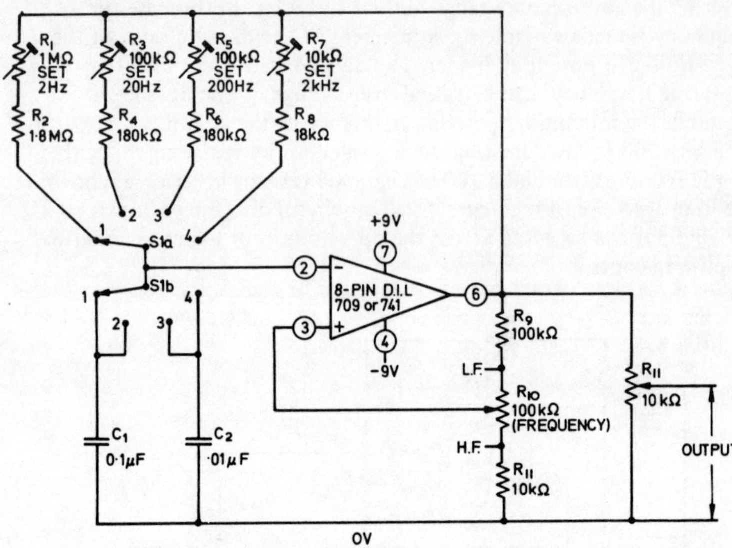

Figure 4.10. 4 decade, 2 Hz - 20 kHz, square wave generator.

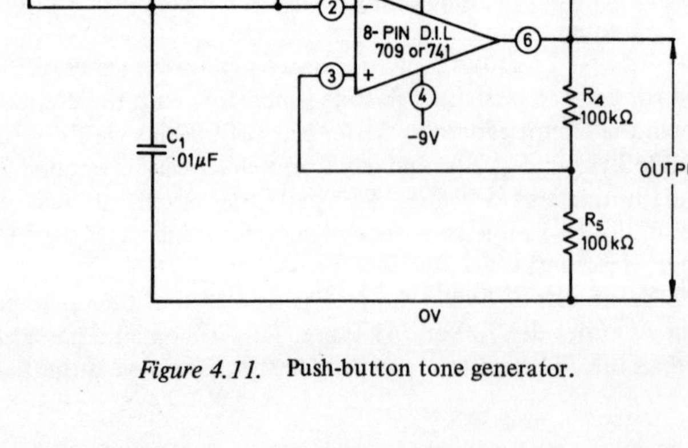

Figure 4.11. Push-button tone generator.

741 gives a good output waveform up to frequencies of about 2 kHz, while the 709 gives a useful waveform up to frequencies of about 20 kHz.

Special waveform generators

The basic relaxation oscillator circuit of *Figure 4.8a* gives a triangle wave output across the timing capacitor, and a square wave output from the op-amp. Both of these waveforms are symetrical, since the positive and negative charging time constants of the circuit are equal. The circuit can be made to give non-symetrical waveforms by using different time constants for the positive and negative half-cycles, as shown in *Figure 4.12*. This circuit gives a rectangular or square wave output that is variable in both frequency and mark/space ratio, the two variable facilities being virtually independent of one another.

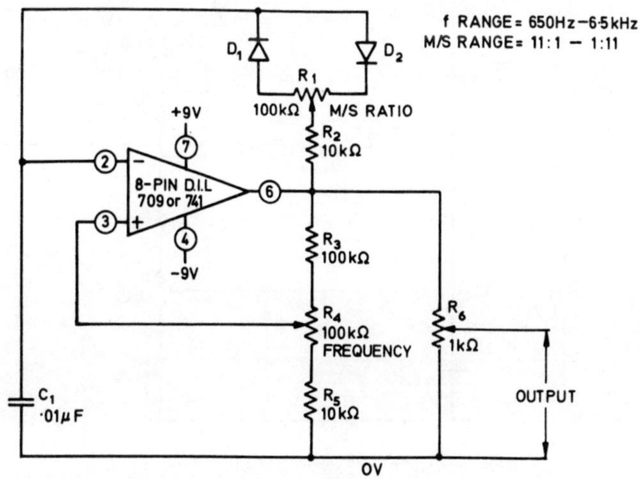

Figure 4.12. Variable frequency and variable m/s ratio, square wave generator.

The variable mark/space ratio facility is obtained via variable pot R_1 and steering diodes D_1 and D_2. On positive charging half-cycles C_1 charges via D_1 and the section of R_1 to the left of the pot slider, and on negative half-cycles it charges via D_2 and the section of R_1 to the right of the slider. Thus, the two time constants of the circuit, and thus the mark/space ratio of the output waveform, are variable via R_1. Note, however, that the sum of the two time constants is also constant, so R_1 has no appreciable effect on the operating frequency of the circuit: the frequency is independently variable via R_4. In practice, variation of the mark/space ratio is restricted to the range 11 : 1 to 1 : 11 via R_1, and

R_1 causes less than 5 % variation in frequency when it is swept through its full range.

With the component values shown, the prototype circuit covers the approximate frequency range 650 Hz to 6.5 kHz. If required, alternative frequency ranges can be obtained by suitable choice of the C_1 values.

Figure 4.13 shows how the *Figure 4.12* circuit can be modified for use as a fixed-frequency (300 Hz) variable-slope ramp generator. The peak-to-peak amplitude of the output ramp is limited to about 1.7 V via the $R_3 - R_4 - R_5$ potential divider to ensure that the waveform is derived from a reasonably linear section of the exponential charging waveform of C_1. R_4 enables the frequency to be set to precisely 300 Hz, although alternative frequencies can, of course, be obtained by using suitable values of C_1.

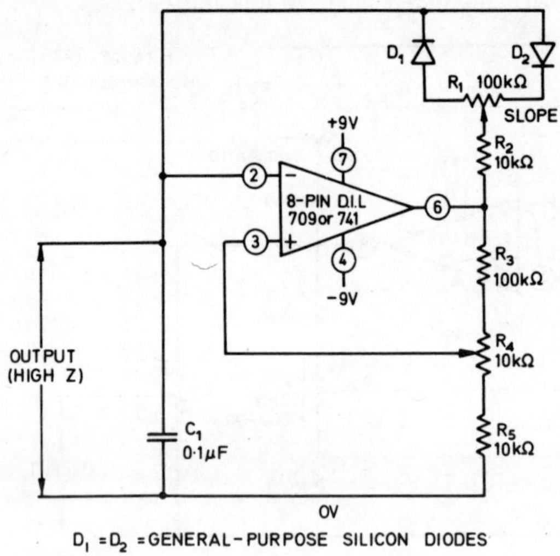

Figure 4.13. 300 Hz variable-slope ramp generator.

The slopes of the output waveform are controlled by R_1. The period of the complete waveform is fixed at 3.33 ms, and R_1 enables the rising (or falling) part of the slope to be varied from approximately 0.3 ms to 3 ms. Note that this circuit has a high output impedance, so the output should be taken to low impedance loads via a suitable buffer stage, such as a unity-gain non-inverting operational amplifier or voltage follower.

The basic relaxation oscillator circuit of *Figure 4.8a* generates a ramp waveform directly from timing capacitor C_1. Since this capacitor is charged exponentially, the ramp output waveform is inevitably slightly non-linear, the degree of non-linearity being proportional to the amplitude of the output signal. A far better system of generating a ramp waveform is shown in *Figure 4.14*. This circuit generates excellent ramp and square waveforms over the frequency range 100 Hz to 1 kHz. The frequency range can be extended by using alternative values of C_1.

The *Figure 4.14* circuit is made up of two sections, these being an integrator (IC_1) and a differential voltage comparator switch (IC_2). To understand the operation of the circuit, assume initially that C_1 is fully discharged and that IC_2 has just switched into the positive saturation state. Under this condition a positive charging voltage is applied to integrating network $R_3 - R_4 - C_1$ via variable potential divider $R_1 - R_2$, so a negative-going linear ramp begins to be generated at the output of IC_1, and is fed to the non-inverting input terminal of IC_2 via R_7. Simultaneously, the positive saturation output voltage of IC_2 is also fed to the IC_2 non-inverting input terminal via R_8.

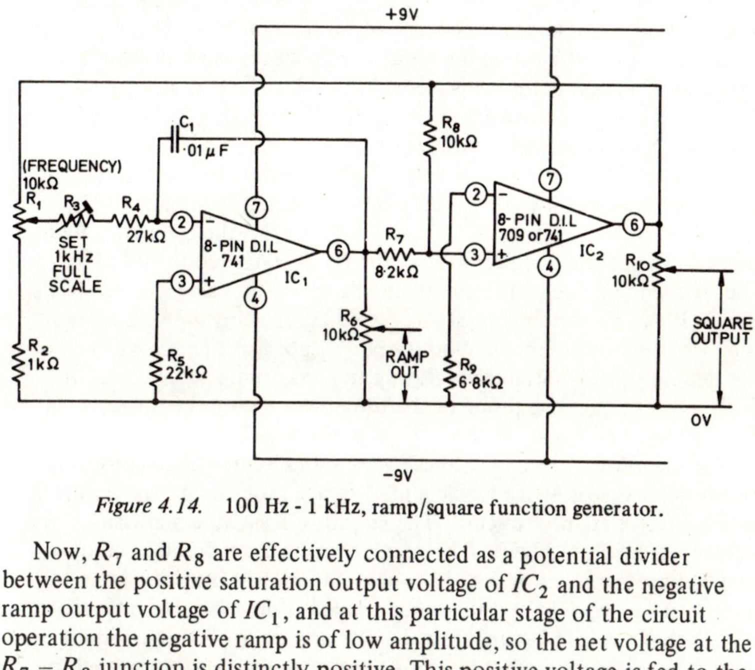

Figure 4.14. 100 Hz - 1 kHz, ramp/square function generator.

Now, R_7 and R_8 are effectively connected as a potential divider between the positive saturation output voltage of IC_2 and the negative ramp output voltage of IC_1, and at this particular stage of the circuit operation the negative ramp is of low amplitude, so the net voltage at the $R_7 - R_8$ junction is distinctly positive. This positive voltage is fed to the non-inverting input of IC_2, which is effectively connected as a non-inverting open-loop d.c. amplifier, so the output of IC_2 is driven to saturation under this condition, as we have already seen.

As time passes, the magnitude of the negative output ramp of IC_1 steadily increases, and the voltage at the $R_7 - R_8$ junction becomes steadily less positive, until eventually the input voltage to the non-inverting terminal of IC_2 falls to zero. At this point IC_2 comes out of saturation and its input begins to drop towards zero. This downward voltage swing is fed back to the input of IC_2 via R_8, and an abrupt regenerative action takes place in which IC_2 switches sharply into negative saturation.

As IC_2 goes into negative saturation the charging voltages on the $R_3 - R_4 - C_1$ integrating network is reversed, so a linear positive-going ramp begins to be generated at the output of IC_1, and simultaneously the voltage at the top of R_8 is switched to a negative saturation value. Thus, the voltage of the $R_7 - R_8$ junction becomes distinctly negative under this condition, and IC_2 is held in negative saturation.

As time passes, the magnitude of the positive output ramp of IC_1 steadily increases, and the voltage at the $R_7 - R_8$ junction becomes steadily less negative, until eventually the input voltage to the non-inverting terminal of IC_2 falls to zero. At this point IC_2 again comes out of saturation and its output begins to drop towards zero. This downward voltage swing is fed back to the input of IC_2 via R_8, and an abrupt regenerative action again takes place, but in this instance IC_2 switches sharply back into positive saturation. The whole timing and switching sequence then repeats ad infinitum.

Thus, the *Figure 4.14* circuit generates a linear ramp output from IC_1 and a square wave output from IC_2. The operating frequency of the circuit can be varied from 100 Hz to 1 kHz via R_1, the frequency being greatest when the R_1 slider is set to the $R_1 - R_2$ junction end of the pot. The frequency range of the circuit can be varied by using alternative C_1 values. If the frequency range is increased appreciably above 2 kHz, a type 709 op-amp should be used in the IC_2 position. The magnitude of the ramp output is variable via R_6, up to a maximum value of about 11 V peak-to-peak. The magnitude of the square-wave output is variable via R_{10}, up to a maximum of about 16 V peak-to-peak.

The linear output ramp of the *Figure 4.14* circuit can be converted into a sine-wave or into a variable mark/space ratio square-wave with the aid of suitable adaptor circuits. The circuit of a variable mark/space ratio adaptor is shown in *Figure 4.15.* Here, the op-amp is wired as an open-loop voltage comparator, and has one input applied from the ramp output of *Figure 4.14* and the other applied from a variable potential divider that is wired between the positive and negative voltage supply lines. The op-amp switches into positive or negative saturation each time that the ramp voltage goes more than a few millivolts below or above the reference voltage set on the non-inverting input via R_2. By adjusting the reference voltage, therefore, the op-amp can be made to change state at any point on the ramp waveform, and a variable mark/space ratio square-wave is thus available at the output of the op-amp.

The circuit of the sine wave adaptor is shown in *Figure 4.16*. Here, the ramp waveform is fed into a resistor-diode matrix via adjustable

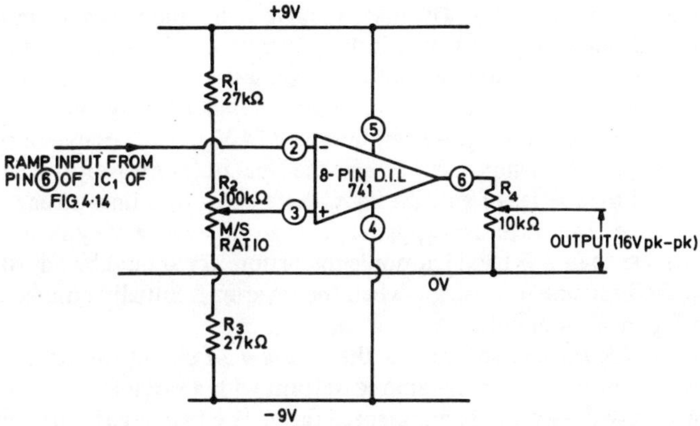

Figure 4.15. Variable m/s-ratio adaptor for function generator (see *Figure 4.14*).

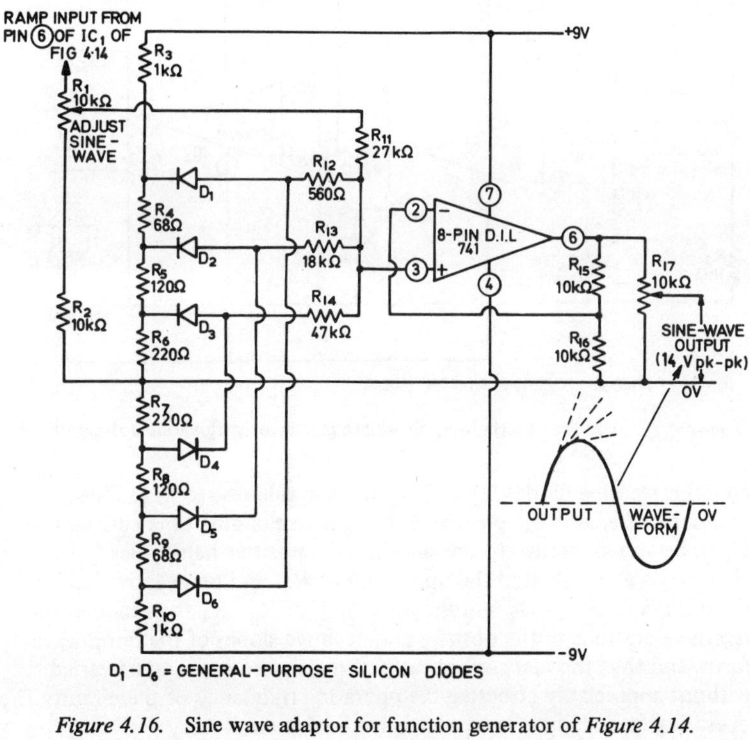

Figure 4.16. Sine wave adaptor for function generator of *Figure 4.14*.

potential divider $R_1 - R_2$. The matrix converts the ramp waveform into a simulated sine wave by automatically reducing the slope of the ramp in a series of steps as the ramp amplitude increases. The resulting waveform is fed into a x 2.2 non-inverting d.c. amplifier, and is finally available with a maximum peak-to-peak amplitude of 14 V across variable output control R_{17}. As shown in the diagram, the final output waveform can be represented by a series of straight lines, there being four lines to each quarter cycle. The waveform approximates a sine wave, and typically contains less than 2 % total harmonic distortion: R_1 should be adjusted to give the best sine wave shape when the inverter is initially connected to the *Figure 4.14* circuit.

Finally, *Figure 4.17* shows how the *Figure 4.14* circuit can be modified to produce a linear ramp waveform with a variable slope, or a square wave with a variable mark/space ratio. The two circuits are similar, except that the integrator charging network of the *Figure 4.17* circuit

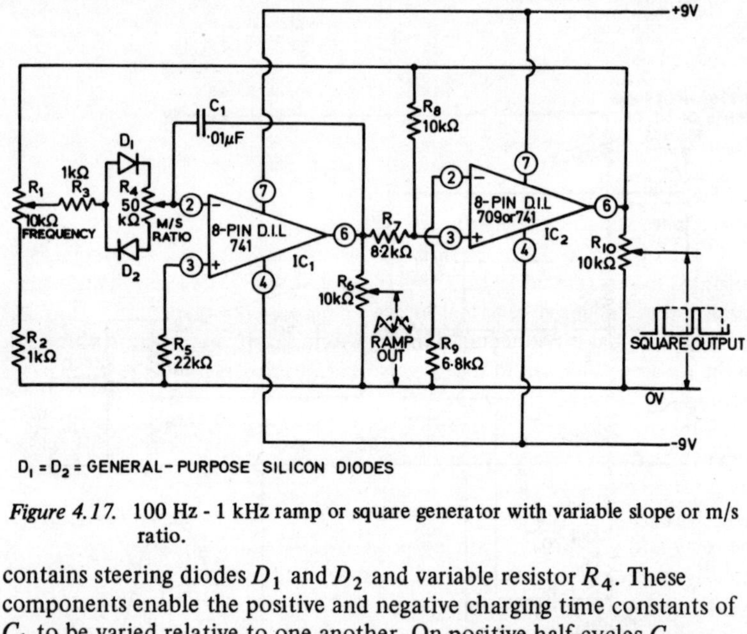

Figure 4.17. **100 Hz - 1 kHz ramp or square generator with variable slope or m/s ratio.**

contains steering diodes D_1 and D_2 and variable resistor R_4. These components enable the positive and negative charging time constants of C_1 to be varied relative to one another. On positive half-cycles C_1 charges via $R_3 - D_1$ and the upper half of R_4, and on negative half-cycles C_1 charges via $R_3 - D_2$ and the lower half of R_4. R_4 thus enables the relative durations of the positive and negative slopes of the ramp waveform, and thus the mark/space ratio of the square wave, to be varied without appreciably effecting the operating frequency of the circuit: The operating frequency is independently variable via R_1.

Multivibrator circuits

Operational amplifiers can usefully be employed in a number of multivibrator applications. *Figure 4.18*, for example, shows how an op-amp can be connected to form a simple bistable multivibrator that is used to give push-button ON-OFF operation of a relay, or of a lamp or similar resistive load.

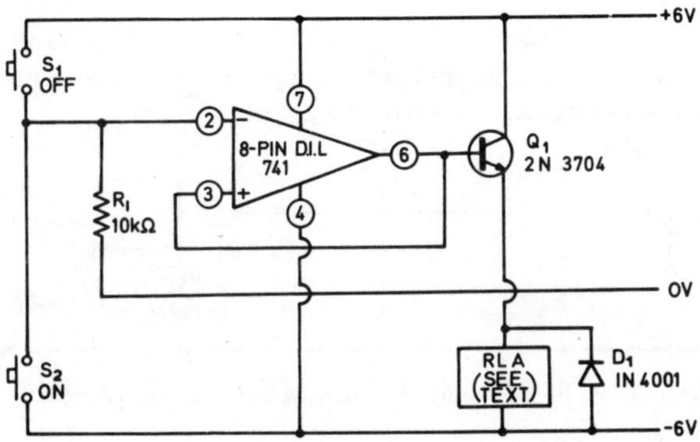

Figure 4.18. Simple manually triggered bistable multivibrator, with relay output.

The op-amp in this circuit acts effectively as an electronic switch, its output being in either a positive or negative saturation state. To understand the operation of the circuit, assume initially that the op-amp output is in positive saturation, and S_1 and S_2 are both open. Under this condition the negative terminal of the op-amp is grounded via R_1, and the positive terminal is connected directly to the positive saturation voltage, so the op-amp is locked in the positive saturation state and the relay is driven on by emitter follower Q_1.

Suppose now that S_1 is momentarily closed, briefly shorting the negative terminal directly to the positive supply line. Under this condition the positive terminal momentarily becomes negative relative to the negative terminal, so the op-amp switches into negative saturation, and the relay and Q_1 turn off. Since the positive terminal is connected directly to the op-amp output, the op-amp automatically locks into the negative saturation state, and remains locked once S_1 is released.

Finally, suppose that S_2 is now momentarily closed, briefly shorting the negative terminal to the negative line. Under this condition the positive terminal momentarily becomes positive relative to the negative terminal, so the op-amp switches back into positive saturation and turns the relay on again, thus completing the cycle of events. The relay can thus be turned on by briefly closing S_2, and can be turned off by briefly closing S_1.

Note that the *Figure 4.18* circuit makes use of two 6 V supply sources. The circuit can be modified for operation from a single 12 V supply, if required, by using the connections shown in *Figure 4.19*. In this case R_2 and R_3 act simply as a potential divider across the supply to give a common reference potential of 6 V to the negative terminal of the op-amp.

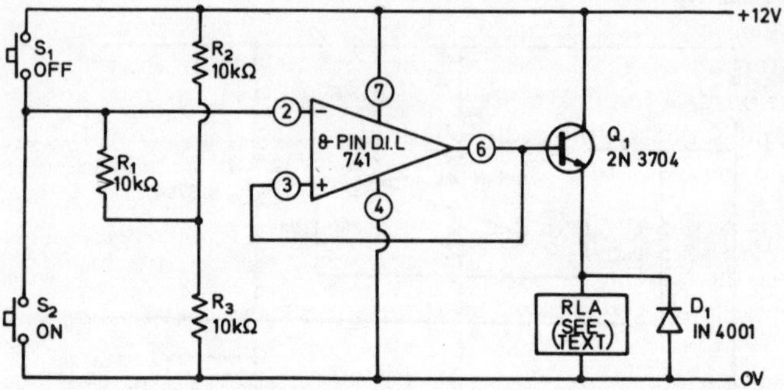

Figure 4.19. Single supply relay-driving bistable multivibrator.

Finally, *Figure 4.20* shows the circuit of a simple manually-triggered relay-driving monostable multivibrator. To understand the circuit operation, assume initially that S_1 is open, and the circuit is in a quiescent state. Under this condition the positive terminal is grounded via R_3, and the negative terminal is held at a small positive potential via $R_1 - R_2$, so the op-amp output is in a state of negative saturation, and the relay is held off via emitter follower Q_1. C_1 is charged up in such a way that its R_3-end is several volts positive relative to its Q_1-base end.

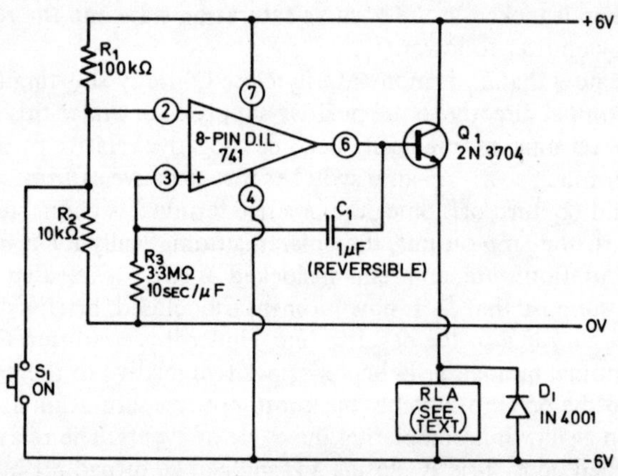

Figure 4.20. Simple 10 s relay-driving monostable multivibrator.

Suppose now that S_1 is momentarily closed, briefly connecting the negative terminal of the op-amp to the negative supply line. Under this condition the negative terminal momentarily becomes negative relative to the positive terminal, so the op-amp output swings into positive saturation and drives the relay on via Q_1. As the op-amp output swings into positive saturation it carries the C_1 charge with it, thus forcing the positive terminal to swing to a positive potential of approximately 10 V, thereby ensuring that the op-amp is locked in the positive saturation state even when S_1 is released. C_1 then slowly starts to discharge via R_3, and the positive terminal voltage decays slowly towards zero. Eventually, after a pre-set delay, the positive terminal voltage falls below the negative terminal voltage (assuming that S_1 is open at this point), and at this stage the op-amp again switches back into negative saturation and turns the relay off. The operating sequence of the circuit is then complete.

Thus, the relay in the *Figure 4.20* circuit turns on as soon as S_1 is momentarily operated, but turns off again automatically after a pre-set delay. The circuit gives a delay of approximately 10 s per microfarad of C_1 value, and thus gives a delay of 10 s with the component values shown. Note that C_1 must be a non-polarised type of capacitor.

The relays used in the *Figures 4.18* to *4.20* circuits can be of any 12 V types with coil resistances greater than 180 ohms or so. If required, resistive loads such as lamps can be used as loads in place of the relays.

CHAPTER 5

20 SOUND GENERATOR AND ALARM PROJECTS

Operational amplifiers can be used in a variety of sound generator and audible-alarm applications. They can readily be adapted for use as code-practice oscillators, door signalling systems, simple musical instruments, metronomes, audible water alarms, light or heat-activated alarms, and as contact-operated audible alarms, etc. In such applications the op-amps can be used to give output powers of a few milliwatts directly, or of several watts with the aid of additional power-booster circuitry.

Twenty useful op-amp sound generator and audible-alarm projects of various types are described in the present chapter. All of these projects are designed around the popular type 741 integrated-circuit op-amp, and the pin connections shown in the following diagrams apply to the 8-pin dual-in-line version of the device only.

Basic relaxation oscillator circuits

Figure 5.1a shows the basic circuit of a simple relaxation oscillator or square-wave generator of the type described in Chapter 4. The op-amp in this circuit is used as a combined voltage comparator and regenerative switch, and the circuit action is as follows.

Suppose initially that the op-amp has just regenerated into positive saturation. Under this condition half of the saturation voltage is switched to the positive terminal via potential divider $R_2 - R_3$, and a rising exponential voltage is switched to the negative terminal via the $R_1 - C_1$ time-constant network. Since the positive terminal is positive relative to the negative terminal at this moment, the op-amp is held in positive saturation. As time passes C_1 charges up via R_1, and the voltage at the negative terminal rises exponentially towards the saturation level. Eventually, a point is reached at which the negative terminal voltage becomes slightly more positive than that on the positive terminal, and the op-amp comes out of saturation. As the op-amp starts to come out of positive saturation a regenerative action is initiated, and the op-amp switches abruptly into negative saturation.

When regeneration is complete half of the negative saturation voltage is applied to the positive terminal via $R_2 - R_3$, and a negative rising exponential voltage is applied to the negative terminal via $R_1 - C_1$. The terminal polarities ensure that the op-amp is held in negative saturation at this time. As time passes the negative terminal becomes progressively

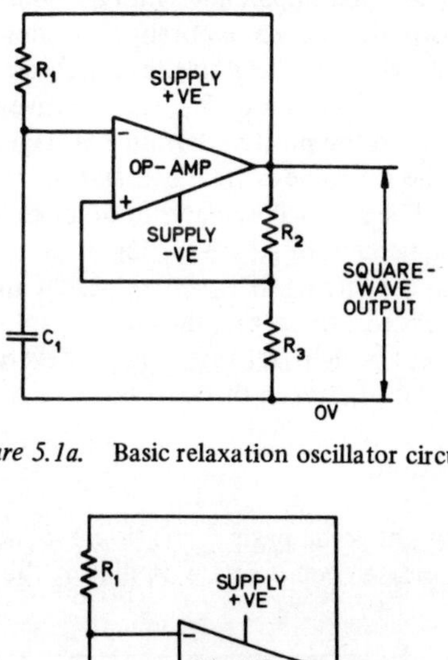

Figure 5.1a. Basic relaxation oscillator circuit.

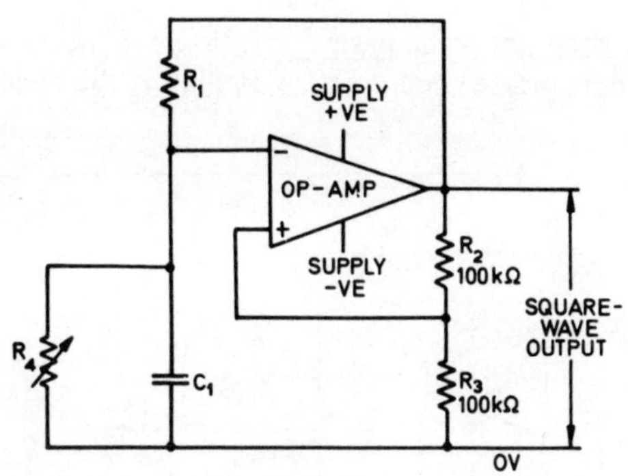

Figure 5.1b. Basic resistance-activated relaxation oscillator circuit.

more negative, and eventually becomes slightly more negative than the positive terminal. At this point a regenerative action is again initiated, and the op-amp switches back into positive saturation. The sequence then repeats ad infinitum.

Thus, the *Figure 5.1a* circuit generates a square wave output, and has its frequency controlled by $R_1 - R_2 - R_3$ and C_1. Note that the frequency can be altered by changing any one of these component values, and that oscillation will cease if R_1 is removed from the circuit. Also note from the description of circuit operation above that the op-amp actually changes state in each half-cycle at the point at which the voltage on the negative terminal just exceeds the voltage on the positive terminal. It follows, therefore, that if the negative terminal is prevented from rising to the positive terminal potential, the circuit will cease to oscillate. This simple fact is utilised in the resistance-activated relaxation oscillator circuit of *Figure 5.1b*.

The *Figure 5.1b* circuit is identical to that of *Figure 5.1a*, except that variable resistor R_4 is wired in parallel with C_1 and in series with R_1: the $R_1 - R_4$ combination thus acts as a potential divider that limits the maximum voltage to which the negative terminal can rise. Consequently, if the R_4 value is lower than that of R_1, the negative terminal voltage will be unable to exceed the positive terminal voltage, and the circuit will fail to oscillate. If the R_4 value is greater than that of R_1, on the other hand, the circuit will be able to oscillate. In practice, R_4 resistance changes of only one percent or so are sufficient to initiate or inhibit oscillatory action in the circuit. If R_4 is replaced with a thermistor or a light-independent resistor, therefore, the circuit can be made to function as a very sensitive heat or light-activated alarm. Several practical circuits of this type are described later in this chapter.

Code-practice oscillators

Figure 5.2a shows how the basic relaxation oscillator can be adapted for use as a simple morse code-practice oscillator: the code-practice key

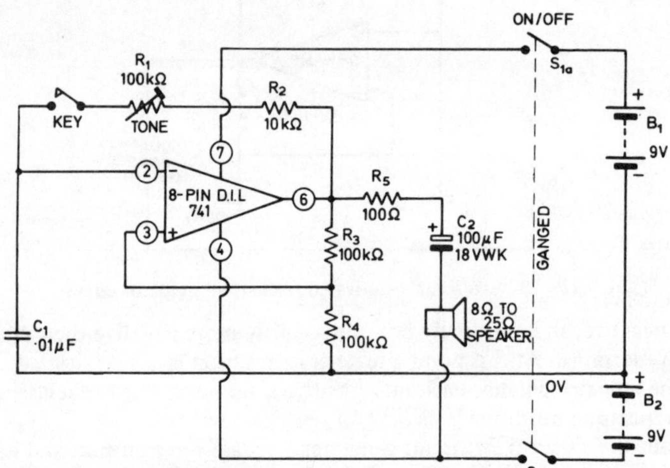

Figure 5.2a. Simple code-practice oscillator.

is wired in series with R_1, so oscillation occurs only when the key is depressed. R_1 is a variable resistor, and enables the practice tone to be varied over the approximate range 400 Hz to 4.4 kHz, to suit individual tastes. The output of the op-amp is fed to a small 8 Ω to 25 Ω speaker via R_5 and C_2. Output powers of only a few milliwatts are generated in the speaker, but this level is more than adequate for this particular application.

The *Figure 5.2a* circuit requires the use of two supply sources. *Figure 5.2b* shows the modifications needed for operation from a single supply source. R_6 and R_7 act as a simple potential divider that enables the positive and negative supply voltages of the op-amp to be derived from a single supply battery, which can be any convenient type that gives a voltage in the range 9 V to 18 V.

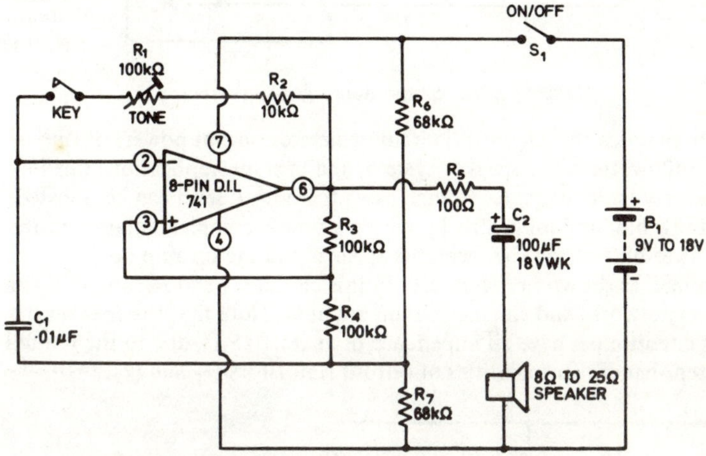

Figure 5.2b. Single supply code-practice oscillator.

Door-signalling systems

Figure 5.3 shows how the basic relaxation oscillator can be adapted for use in a simple door-signalling system. In this case the circuit is provided with three tone-controlling timing resistors, which can be selected via push-button switches S_1 to S_3. With the component values shown a tone of roughly 200 Hz is produced via S_1, 440 Hz is generated via S_2, and 900 Hz is generated via S_3. Note that these tone frequencies rise in steps of roughly one octave, so each push-button switch generates a very distinctive tone. Thus, if S_1 is placed at, say, the front door of the house, S_2 is placed at the side door, and S_3 is placed at the back door, the position of any caller will be readily apparent from the tone of the call signal.

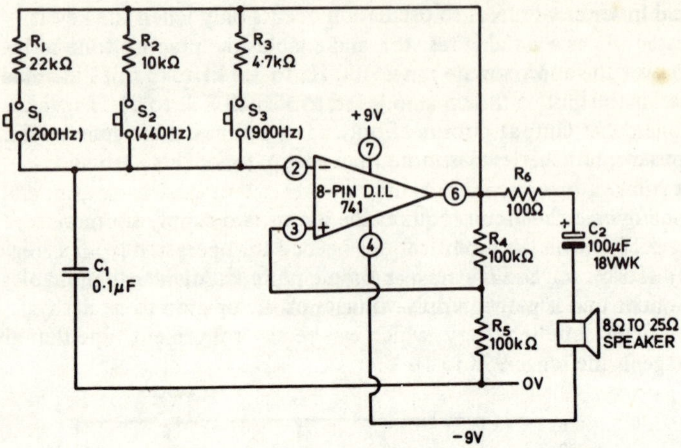

Figure 5.3. Simple door-signalling system.

In practice the *Figure 5.3* circuit generates output powers of only a few milliwatts in the speaker system, and in some applications this low level may be inadequate. In such cases the output level can be boosted to a few hundred milliwatts by wiring a simple complementary emitter-follower buffer stage between the speaker and the op-amp output terminal, as shown in *Figure 5.4*. In this circuit C_3 and R_6 are used as a Zobel network, and enhance circuit stability. Note that the speaker in this circuit must have an impedance of at least 25 Ω, due to the limited current-handling capabilities of output transistors Q_1 and Q_2.

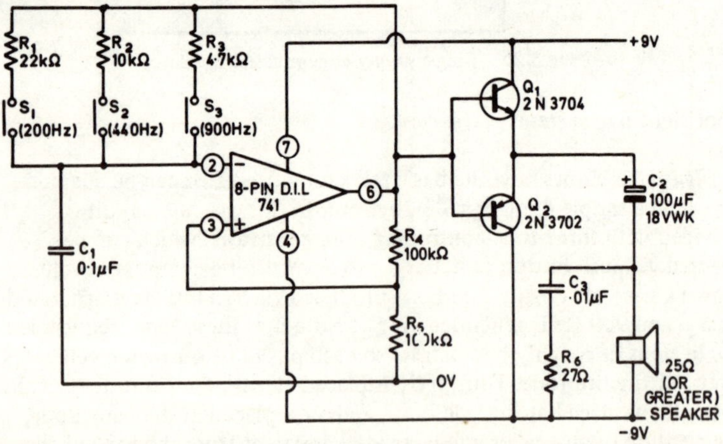

Figure 5.4. Higher-power door-signalling system.

Musical instrument circuits

Figure 5.5 shows how the basic relaxation oscillator can be adapted for use as the basis of a simple keyboard musical instrument. In this case the tone-controlling resistor network consists of a chain of twenty-five resistors wired in series. The junction of each pair of resistors in the chain is taken to a copper or similar low-resistance keyboard strip, and any one of these strips can be connected to the op-amp output via a flying probe. Thus, different tones can be selected by touching the probe onto different keyboard strips, and the circuit functions as a musical instrument that is played via the touch probe.

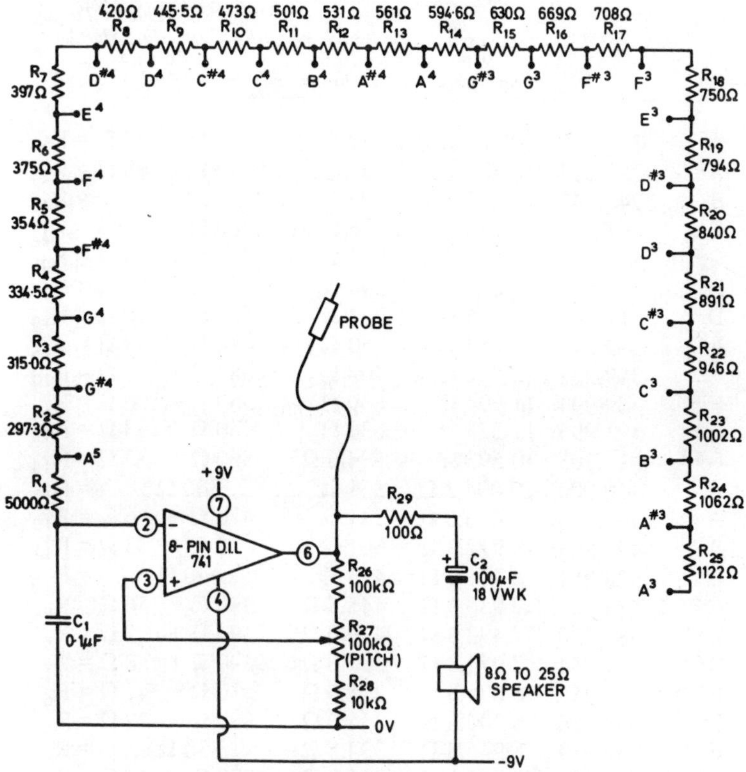

Figure 5.5. A simple keyboard musical instrument.

In practice this simple musical instrument gives 25 notes, and covers two full octaves, from concert pitch A^3 (220 Hz) to A^5 (880 Hz). An adjustable pitch control is included in the circuit, and enables the scale to be shifted by more than a decade, to give a top range of roughly 2 200 Hz to 8 800 Hz. *Table 5.1* shows details of the relationships between resistance, frequency, and notes of the keyboard of the *Figure 5.5* circuit.

The musical instrument circuit of *Figure 5.5* can be improved and modified in a number of ways. *Figure 5.6*, for example, shows how the maximum output power of the circuit can be boosted to a few hundred milliwatts by incorporating a complementary emitter-follower stage

Table 5.1. Relationships between resistance, frequency, and notes of the keyboard of the *Figure 5.5* circuit.

Note	*Frequency (Hz)*	*Sum of the Timing Resistor Chain, in Ohms*	*Value of the Incremental Timing Resistor, in Ohms*	*Practical Value or Timing Resistor, using Standard Resistors*
A^3	220.000	20 109 Ω	1 122 Ω	1 kΩ + 120 Ω = R_{25}
$A\#^3$	233.082	18 787 Ω	1 062 Ω	1 kΩ + 68 Ω = R_{24}
B^3	246.942	17 985 Ω	1 002 Ω	1 kΩ = R_{23}
C^3	261.626	16 923 Ω	946 Ω	860 Ω + 86 Ω = R_{22}
$C\#^3$	277.183	15 977 Ω	891 Ω	860 Ω + 33 Ω = R_{21}
D^3	293.665	15 086 Ω	840 Ω	680 Ω + 150 Ω = R_{20}
$D\#^3$	311.127	14 246 Ω	794 Ω	680 Ω + 120 Ω = R_{19}
E^3	329.628	13 452 Ω	750 Ω	680 Ω + 68 Ω = R_{18}
F^3	349.228	12 702 Ω	708 Ω	680 Ω + 27 Ω = R_{17}
$F\#^3$	369.994	11 994 Ω	669 Ω	560 Ω + 100 Ω = R_{16}
G^3	391.995	11 225 Ω	630 Ω	560 Ω + 68 Ω = R_{15}
$G\#^3$	415.305	10 595 Ω	594.6 Ω	560 Ω + 33 Ω = R_{14}
A^4	440.000	10 004.3 Ω	561 Ω	560 Ω = R_{13}
$A\#^4$	466.164	9 443.0 Ω	531 Ω	470 Ω + 56 Ω = R_{12}
B^4	493.883	8 912.3 Ω	501 Ω	470 Ω + 33 Ω = R_{11}
C^4	523.251	8 411.3 Ω	473 Ω	470 Ω = R_{10}
$C\#^4$	554.365	7 938.3 Ω	445.5 Ω	390 Ω + 56 Ω = R_9
D^4	587.330	7 492.8 Ω	420.0 Ω	390 Ω + 27 Ω = R_8
$D\#^4$	622.254	7 072.8 Ω	397.0 Ω	390 Ω + 6.8 Ω = R_7
E^4	659.255	6 675.8 Ω	375.0 Ω	330 Ω + 47 Ω = R_6
F^4	698.456	6 300.8 Ω	354.0 Ω	330 Ω + 22 Ω = R_5
$F\#^4$	739.989	5 946.8 Ω	334.5 Ω	330 Ω = R_4
G^4	783.991	5 612.3 Ω	315.0 Ω	270 Ω + 47 Ω = R_3
$G\#^4$	830.609	5 297.3 Ω	297.3 Ω	270 Ω + 27 Ω = R_2
A^5	880.000	5 000.0 Ω	5 000 Ω	10 kΩ // 10 kΩ = R_1

NOTE: Incremental ratio of frequency and resistance = 1.0594631 between the speaker and the op-amp output, and how an adjustable volume control can be included in the design.

Figure 5.7 shows how a vibrato facility can be added to the circuit. Here, IC_2 is wired as a Zener-regulated 8 Hz Wien-bridge sine wave oscillator, and its output waveform is used to apply 8 Hz frequency modulation or vibrato to the main oscillator stage of the instrument. The vibrato depth can be controlled by a 10 kΩ pot, as shown.

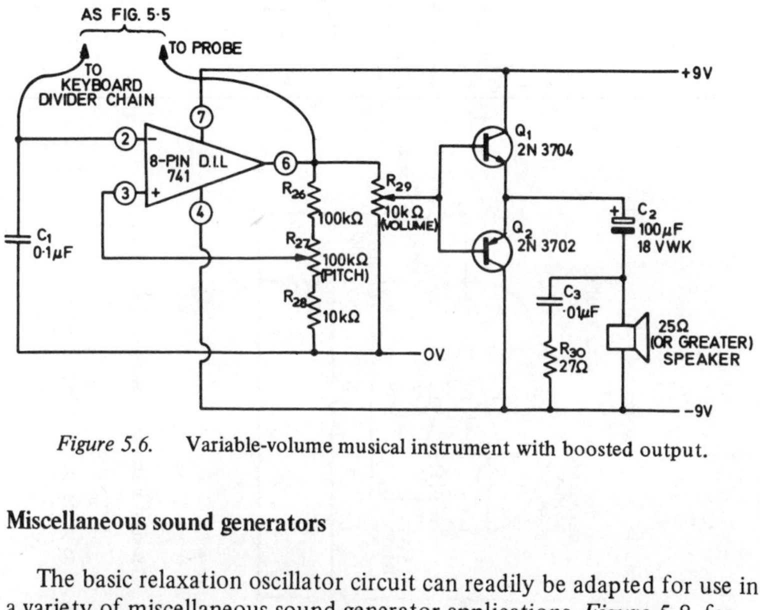

Figure 5.6. Variable-volume musical instrument with boosted output.

Miscellaneous sound generators

The basic relaxation oscillator circuit can readily be adapted for use in a variety of miscellaneous sound generator applications. *Figure 5.8*, for example, shows how it can be made to function as a light-sensitive oscillator or sound generator by simply using a cadmium-sulphide light-dependent resistor (LDR) in the tone-controlling timing resistor position, and *Figure 5.9* shows how it can be made to function as a heat-sensitive oscillator or sound generator by using a thermistor in the same position. Note in the *Figure 5.9* circuit that series resistor R_1 is used to limit the frequency range of the design, and that the frequency range is made variable via R_3. In these two circuits the LDR or thermistor can be any type having a resistance value in the range 500 Ω to 5 MΩ under actual operating conditions, thus enabling the operating frequency to be varied over a 10 000:1 range.

Figure 5.10 shows how the basic relaxation oscillator can be modified for use as a simple metronome, designed to cover the range 30 to 300 beats per minute, and thus encompassing the usual musical tempo range of largo (40 beats per minute) to presto (250 beats per minute). The

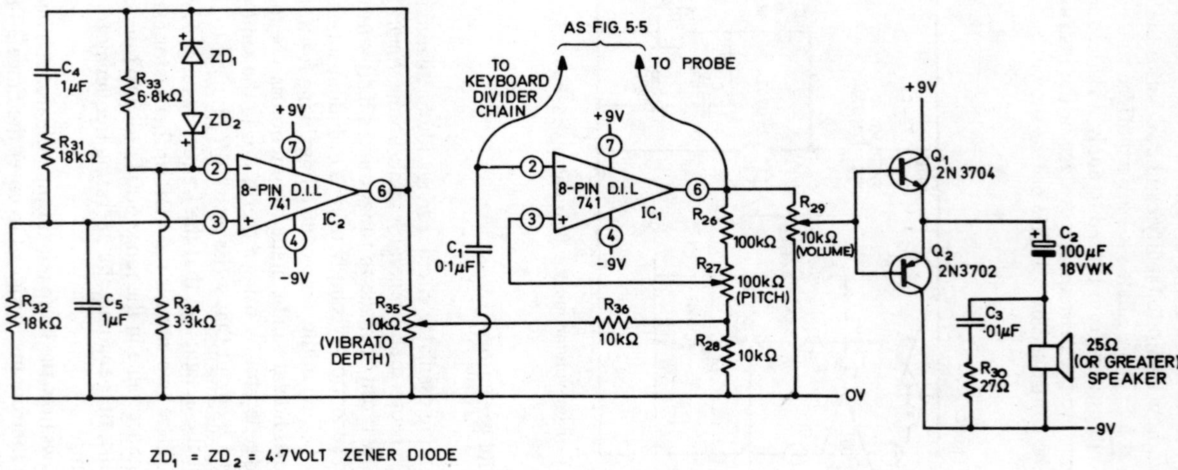

Figure 5.7. Musical instrument with built-in vibrato.

circuit generates a rectangular waveform with a mark/space ratio of approximately 100:1, and thus delivers a brief pulse of energy to the speaker in each operating cycle. This energy pulse causes a double click to be heard in the speaker at periodic intervals.

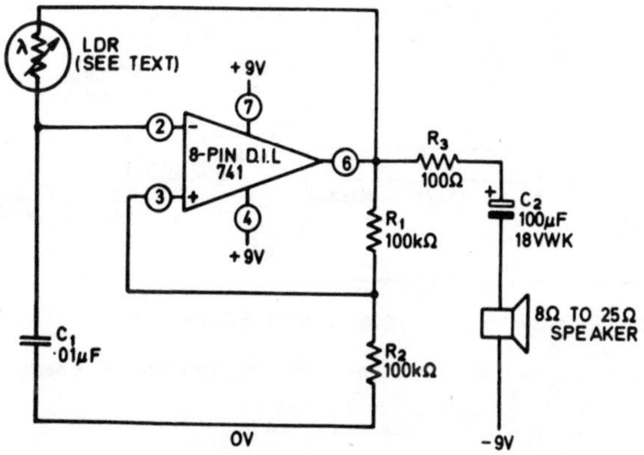

Figure 5.8. Light-sensitive oscillator/sound-generator.

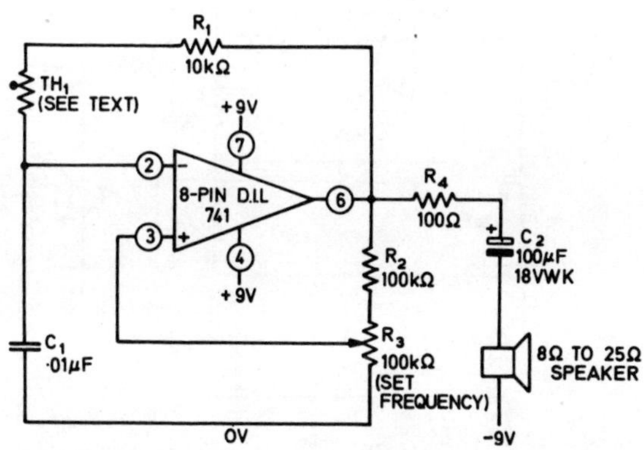

Figure 5.9. Heat-sensitive oscillator/sound-generator, with frequency limiting and adjustable frequency control.

The operating frequency of this circuit is variable via R_5, and the maximum operating period can be precisely pre-set to one beat per two seconds via R_2.

Finally, *Figure 5.11* shows how the relaxation oscillator can be connected for use as a water-activated alarm. In this case a fixed 100 kΩ resistor is used as the tone-controlling element, and is connected to the oscillator via a pair of metal probes. When any reasonably conductive

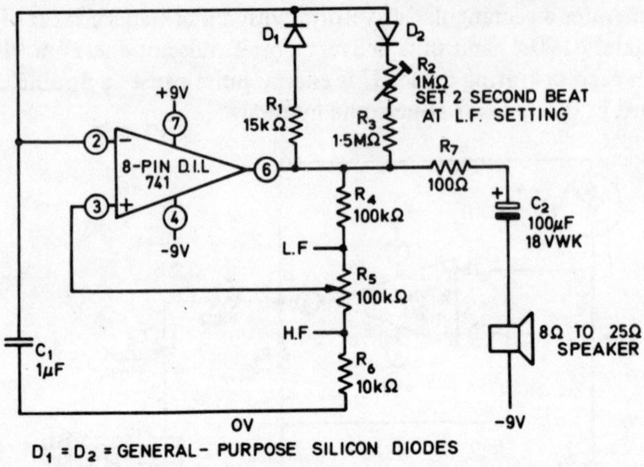

Figure 5.10. Simple metronome, covering 30 to 300 beats per minute.

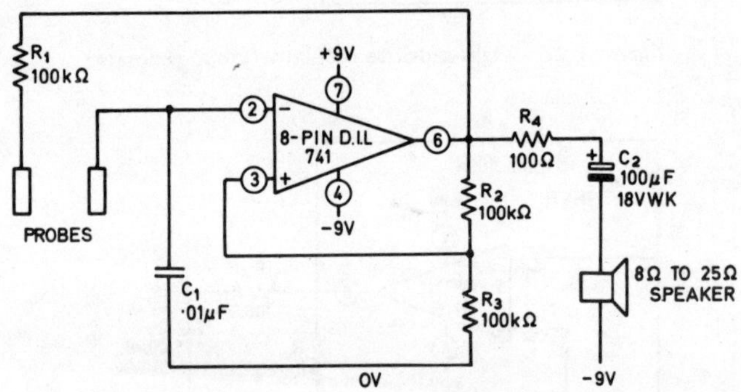

Figure 5.11. Water-activated alarm.

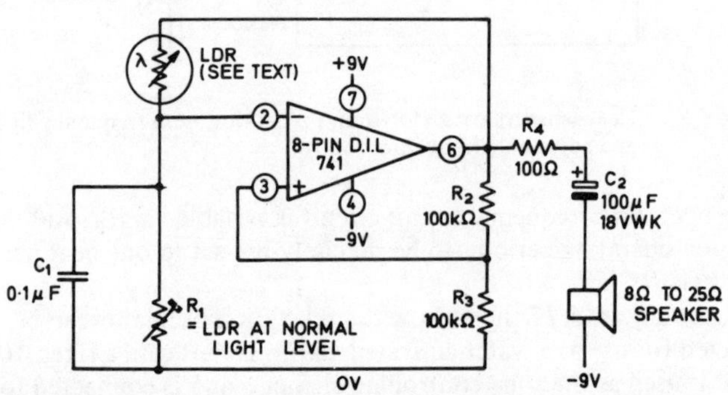

Figure 5.12. Precision light-activated alarm.

liquid comes into contact with both of these probes simultaneously, the circuit oscillates and generates an alarm tone of a few hundred Hz: R_1 limits the upper frequency (with the probes shorted) to about 400 Hz. By using suitably adapted probes, the alarm can be made to sound when water reaches a certain level in a bath or tank, or when flooding occurs in a cellar or basement, or when rain falls across the probes, etc. The circuit draws, typically, a current of less than 1 mA from each supply battery when it is in the standby condition.

Heat and light-activated alarms

It was pointed out earlier in this chapter that the basic relaxation oscillator of *Figure 5.1a* can be made to function as a resistance-activated oscillator by simply wiring a shunt resistor across the circuits main timing capacitor, as shown in *Figure 5.1b*. *Figure 5.12* shows how the resistance-activated oscillator can be adapted for use as a precision light-activated alarm that turns on when the light intensity exceeds a pre-set level.

In this circuit a cadmium-sulphide LDR is used as the designs tone-controlling resistive element, and variable resistor R_1 is wired across C_1. In practice R_1 is adjusted so that oscillation just commences when the light level rises to the required level: under this condition the LDR presents a resistance fractionally less than that of R_1, so the op-amps negative terminal voltage is able to exceed that of the positive terminal and the circuit oscillates. If the light level falls below the pre-set level the LDR resistance increases above that of R_1, and under this condition the negative terminal voltage is unable to exceed that of the positive terminal, so oscillation ceases. The sensitivity of this circuit is so high that the alarm can be turned on and off by changes in light level too small to be detected by the human eye.

The *Figure 5.12* circuit can be made to function as a dark-activated alarm, which turns on when the light intensity falls below a pre-set level, by simply transposing the R_1 and LDR positions, as shown in *Figure 5.13*. This *Figure 5.13* circuit can be made to function as a sensitive smoke alarm by simply illuminating the LDR via a stabilised light beam and adjusting R_1 so that the circuit just fails to oscillate. If smoke is subsequently injected into the light beam the LDR illumination intensity will fall, and the alarm will sound.

The *Figure 5.12* and *5.13* circuits can be used with any cadmium-sulphide LDRs having resistances in the range 2 kΩ to 2MΩ at the required trigger levels. The alarm frequency can be adjusted, if required, by simply changing the C_1 value: increasing the C_1 value reduces the frequency, and vice versa.

Figure 5.14 shows how the *Figure 5.12* circuit can be made to function as a precision over-temperature alarm by simply using a thermistor in place of the LDR. An under temperature alarm can be made by using a thermistor in place of the LDR of the *Figure 5.13* circuit. The thermistors can be any negative temperature coefficient types having a resistance in the range 2 kΩ to 2 MΩ at the required trigger levels.

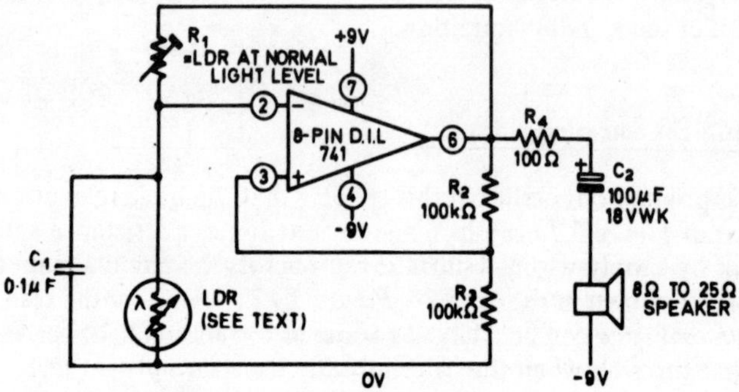

Figure 5.13. Precision dark-activated alarm.

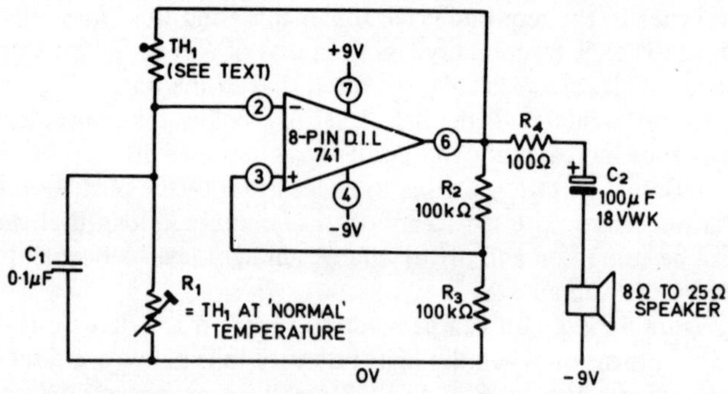

Figure 5.14. Precision over-temperature alarm.

The simple heat and light-activated alarms that we have looked at so far can be modified and adapted in a number of ways to increase their versatility. Their output powers can, for example, be increased to several hundred milliwatts by wiring a simple power-boosted stage between the speaker and the op-amp output terminal, as shown in the high output under-temperature alarm circuit of *Figure 5.15*. Note in this case that the speaker must have an impedance of 25 Ω or greater, due to the limited current-handling capabilities of the specified output transistors.

Again, the circuits can be modified for operation from a single supply battery by using a simple potential divider to give the necessary op-amp

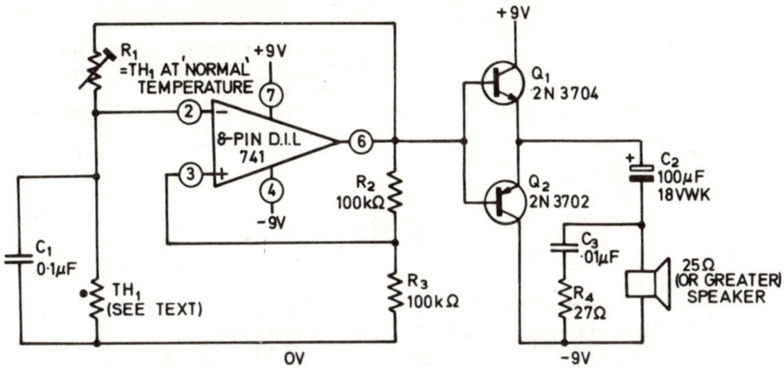

Figure 5.15. High output under-temperature alarm.

biasing, as shown by R_5 and R_6 in the single-supply over-temperature alarm circuit of *Figure 5.16*.

If required, both of the above modifications can be applied to a single alarm system, as shown in the single supply, high output, frost or under-temperature alarm circuit of *Figure 5.17*. This particular design can usefully be built into an automobile, with the thermistor mounted low down on the front of the vehicle, for use as an ice hazard warning alarm.

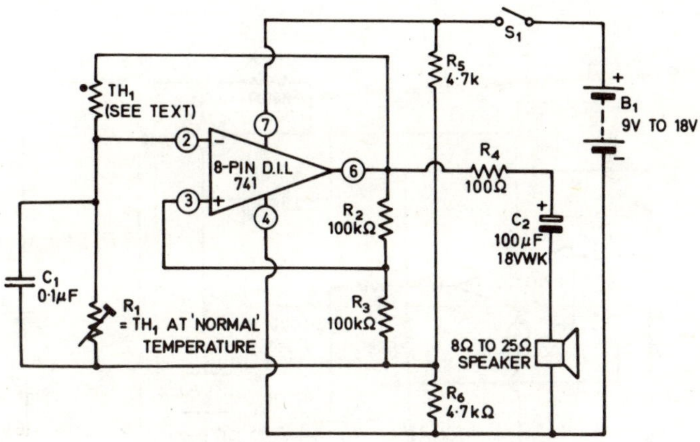

Figure 5.16. Single supply over-temperature alarm.

The two-rail and single-rail light and heat alarm circuits can be made to activate a relay as the alarm sounds by using the connections shown in the frost alarm circuits of *Figures 5.18* and *5.19*. The relay contacts can be used to activate auxiliary warning devices or mechanisms such as lights, heaters, motors, etc.

The operating principle of the relay-driving sections of the *Figure 5.18*

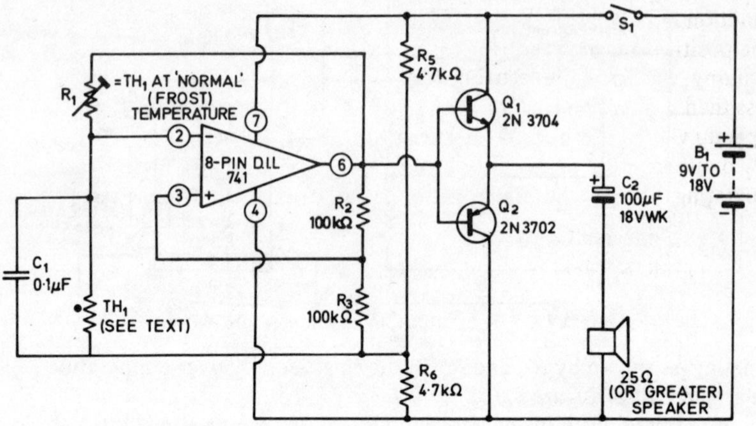

Figure 5.17. Single supply high-output frost alarm.

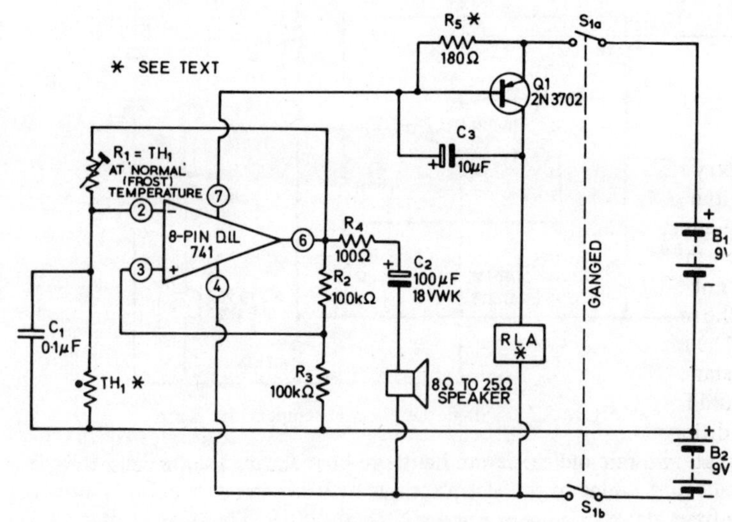

Figure 5.18. Dual supply frost alarm with sound and relay output.

and *5.19* circuits is quite simple. The relay is used as the collector load of Q_1, which is wired as a common emitter amplifier with its base-emitter junction in parallel with 180 Ω resistor R_5, which is wired in series with the positive supply line of the op-amp. Normally, with the alarm on standby, the op-amp draws a quiescent current (from each battery) of less than 1 mA. The current of the positive supply line flows to the op-amp via R_5, so the potential across this resistor is insufficient to bias Q_1 on under this condition, and the relay is off. When the alarm is operating, on the other hand, the circuit current drawn from each

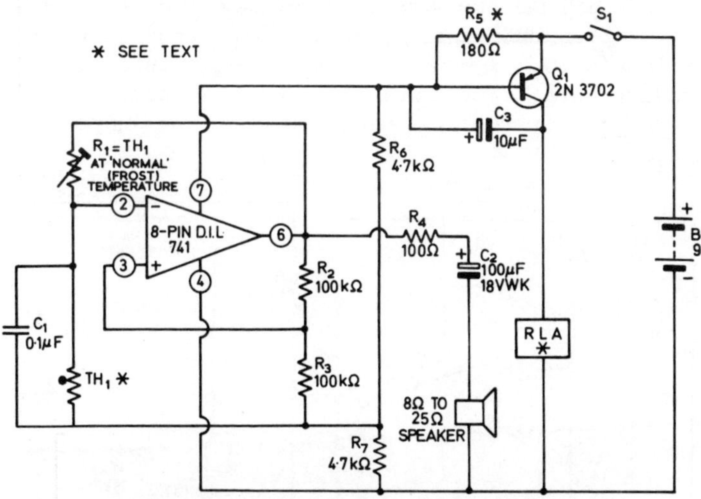

Figure 5.19. **Single supply frost alarm with relay output.**

battery rises to roughly 20 mA, so in this case the R_5 potential drives Q_1 and the relay on, and most of the current reaches the op-amp via the Q_1 base-emitter junction: Q_1 and the relay are thus turned on automatically when the alarm sounds. C_2 is used to differentiate the pulsed collector current of Q_1 under the alarm condition, and converts it to smooth d.c., so the relay operates without chatter.

The relay in each of the above circuits can be any type with a coil resistance greater than 180 Ω, and with a suitable voltage rating. R_5 should have a value low enough not to drive the relay on under quiescent conditions but high enough to turn it on as soon as the alarm starts to oscillate: a value of 180 Ω should prove satisfactory in most cases.

A contact-operated alarm

Finally, to conclude this chapter, *Figure 5.20* shows how the basic op-amp relaxation oscillator circuit can be adapted for use as the basis of a self-latching contact-operated alarm system that can be used as a burglar alarm in the home. In this circuit the op-amp oscillator is adapted for single-supply operation via the $R_4 - R_5$ potential divider, and the complete oscillator is connected in parallel with the coil of relay *RLA*, which is used as the collector load of transistor Q_1. Q_1 is wired as a basic common emitter amplifier, with its base bias derived from the positive supply line via switches S_1 to S_4 and via resistor R_7. The output of the op-amp oscillator is fed to the 25 Ω speaker via power-boosting transistors Q_2 and Q_3. Circuit operation is as follows.

Normally, with the alarm system in the standby condition, switches S_1 to S_3 are open. Under this condition zero bias is applied to the base of transistor Q_1, so the transistor is cut off. Since Q_1 is cut off, zero current flows through the op-amp oscillator circuit, so the alarm is inoperative under this condition, and the circuit consumes only a negligible leakage current.

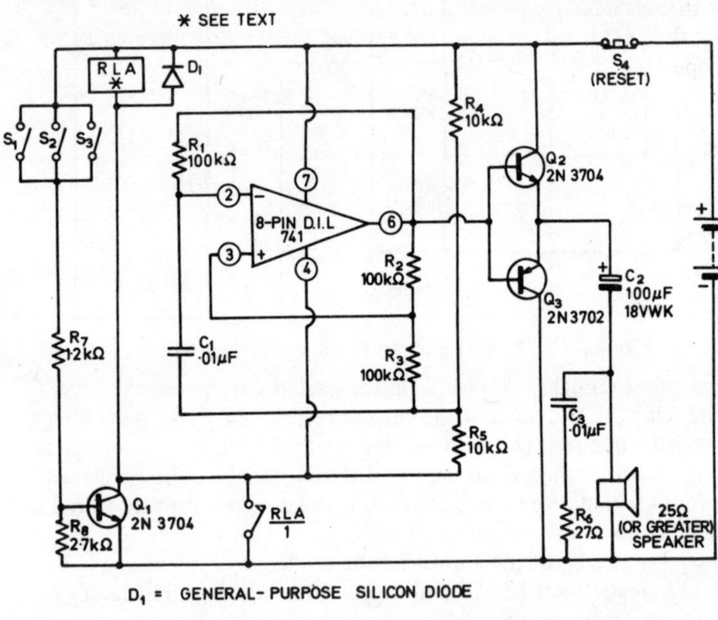

Figure 5.20. Make-to-operate self-latching alarm system.

Suppose that one of the S_1 to S_3 switches is now momentarily closed. As the switch closes it applies base bias to Q_1, and drives the transistor to saturation. As the transistor goes into saturation it operates the relay and causes contacts *RLA*/1 to close, and these contacts then hold the relay in the self-latched state even if switches S_1 to S_3 are subsequently opened. Simultaneously, as power is applied to the relay, power is applied across the op-amp oscillator section of the circuit, and a powerful alarm tone is generated in the speaker. Thus, the alarm operates and self-latches into the ON mode as soon as one or more of the S_1 to S_3 switches is momentarily closed. Once the alarm has latched on, it can be turned off again by operating RESET switch S_4.

In practice, the above circuit can be made to function as a burglar alarm by using microswitches or pressure-pad switches in the S_1 to S_3 positions. The microswitches can be activated by the action of doors or windows opening, and the pressure pads can be activated by the action of stepping onto rugs or carpets, etc. If required, relay *RLA* can be provided with an extra pair of contacts, which can be used to operate auxiliary circuitry such as delayed-action alarm bells, lights, etc. The relay can be any type having a coil resistance greater than 180 Ω and having a voltage rating suitable to the supply voltages used in the circuit. These voltages can have any convenient value in the range 9 V to 18 V. Diode D_1, wired across the relay coil, is a general purpose silicon diode, and is used to protect the circuit against any back e.m.f. that may be generated as the relay operates.

CHAPTER 6

20 RELAY-DRIVING SWITCHING PROJECTS

Operational amplifiers can be used in a variety of high-precision relay-driving switching applications, in which the relay is used to activate external circuitry via its contacts. In these applications the op-amps can readily be made to function as high-precision over-temperature or under-temperature switches, as sensitive light-activated switches, as water or steam-activated switches, as precision a.c. or d.c. voltage-sensitive switches, as time switches, or as switches that are activated by touch, sound, or tone.

Twenty useful and versatile op-amp relay-driving switching projects of various types are described in this final chapter of this volume. All of these projects are designed around the popular type 741 integrated-circuit op-amp, and the pin connections shown in the following diagrams apply to the 8-pin dual-in-line version of the device only.

Precision temperature-activated switches

Figure 6.1 shows how an op-amp can be connected as a relay-driving precision frost or under-temperature switch. In essence, the circuit consists of a fixed potential divider (R_2 and R_3), which generates a stable reference voltage, and a variable potential divider (R_1 and TH_1), which generates a temperature-dependent voltage. These two potential dividers are effectively connected as a Wheatstone bridge, with the bridge output feeding to the input terminals of the operational amplifier. The op-amp is connected as an open-loop differential amplifier or voltage comparator, and its action is such that its output is driven to positive saturation if its negative terminal is more than a few hundred microvolts negative to the positive terminal, and is driven to negative saturation if its negative terminal is more than a few hundred microvolts positive to the positive terminal. The output of the op-amp is coupled to relay-driving common-emitter amplifier Q_1 in such a way that the relay and transistor are cut off when the op-amp is positively saturated, and are driven fully on when the op-amp is negatively saturated.

TH_1 in this circuit is a negative-temperature-coefficient thermistor, so its resistance falls as temperature rises, and vice versa. In practice, variable resistor R_1 is adjusted so that the $R_1 - R_2 - R_3 - TH_1$ bridge is balanced at a temperature very close to the required trip or switching value, and under this condition zero voltage difference exists between the negative and positive terminals of the op-amp. Consequently, when the

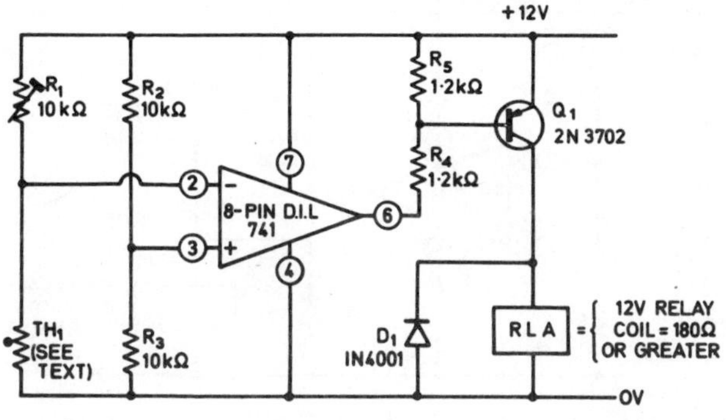

Figure 6.1. Precision frost or under-temperature switch.

temperature rises above the trip level the bridge goes out of balance in such a way that the negative terminal of the op-amp goes negative to the positive terminal, and the op-amp is driven to positive saturation, and Q_1 and the relay are cut off. When the temperature falls below the trip level, on the other hand, the bridge goes out of balance in such a way that the negative terminal of the op-amp goes positive to the positive terminal, and the op-amp is driven to negative saturation and drives Q_1 and the

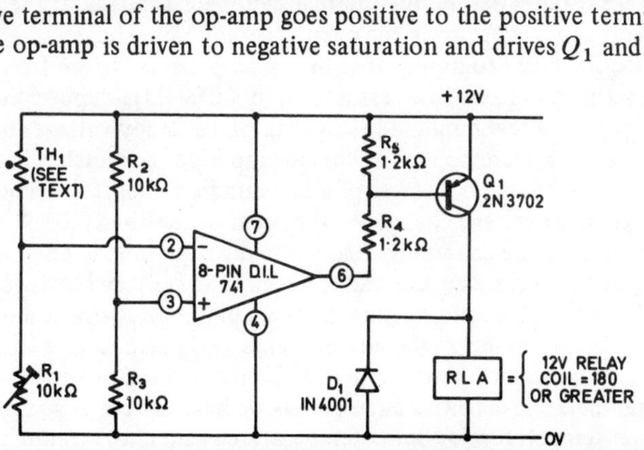

Figure 6.2a. Precision fire or over-temperature switch.

relay on. Thus, the relay turns on whenever the temperature falls below the pre-set trip level.

The action of the above circuit can be reversed, so that the relay turns on whenever the temperature exceeds a pre-set trip level, by simply transposing the R_1 and TH_1 positions, as shown in *Figure 6.2a*. Alternatively, the action can be reversed by leaving R_1 and TH_1 as they are but transposing the negative and positive input terminal connections of the op-amp, as shown in *Figure 6.2b*.

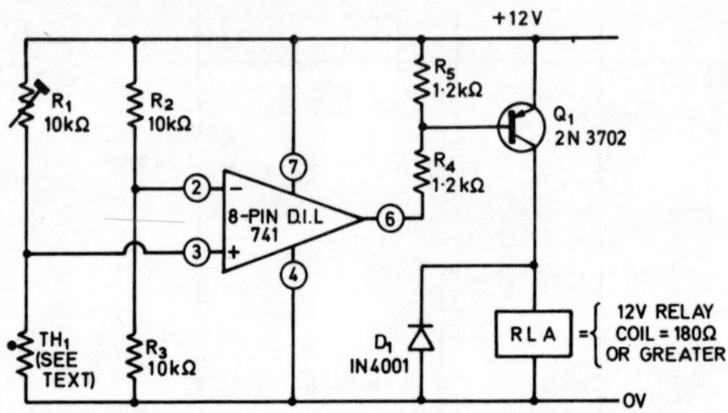

Figure 6.2b. **Alternative precision fire or over-temperature switch.**

Two points of particular importance should be noted concerning the *Figure 6.1* and *6.2* circuits. The first point is that each circuit goes through its switching phase when the bridge is approximately balanced, and the balancing point of a bridge is independent of the actual supply line voltage. Consequently, the thermal switching accuracy of the circuit is unaffected by variations in the circuits supply rail voltages.

The second point to note is that the op-amp can be driven from positive saturation to negative saturation by differential input voltage changes of only a few hundred microvolts. Since steady voltages of approximately 6 V are applied to the op-amp input terminals when the bridge is balanced, these changes of a few hundred microvolts represent percentage input voltage changes of the order of less than 0.01 % and these changes can be caused by a similar percentage shift in any one of the bridge resistances. Consequently, the circuits are very sensitive to changes in the TH_1 resistance, and give true precision temperature-sensing action. In practice, the circuits can be expected to give thermal switching accuracies of better than 0.05 °C at room temperatures.

The temperature-sensitive circuits that we have looked at so far give either over-temperature or under-temperature switching. If required, both types of action can be combined in a single two-relay circuit by using the connections shown in *Figure 6.3*. In this case the left (under-temperature) half of the circuit is based on that of *Figure 6.1*, and the right (over-temperature) half is based on *Figure 6.2b*. Both halves of the circuit share a common $R_1 - TH_1$ temperature-sensing section, but the under-temperature and over-temperature switching levels of the circuit are independently adjustable; the setting up procedure is as follows.

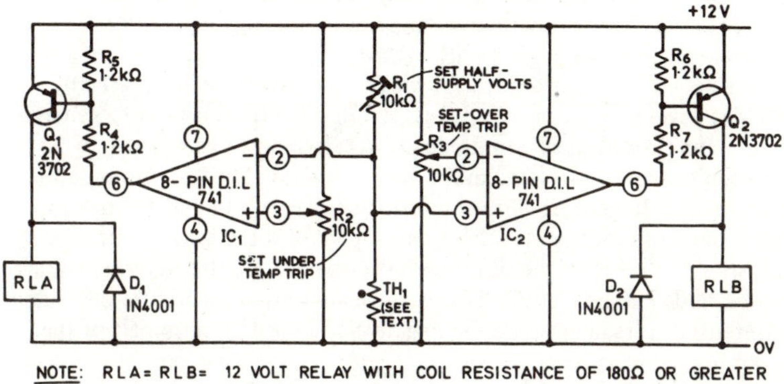

Figure 6.3. Combined under/over-temperature switch, with independent outputs.

First, set R_2 and R_3 to approximately mid travel, and then, with thermistor TH_1 at its 'normal' temperature, adjust R_1 so that approximately half-supply volts are developed across TH_1. Now reduce TH_1 to the required under-temperature trip level, and adjust R_2 so that relay *RLA* just turns on. Finally, raise TH_1 to the required over-temperature trip level, checking that *RLA* turns off, and adjust R_3 so that relay *RLB* just turns on. All adjustments are then complete, and the circuit is ready for use.

The thermistors used in the *Figure 6.1* to *6.3* circuits can be any negative-temperature-coefficient type that presents a resistance in the approximate range 900 Ω to 9 000 Ω at the required trip temperatures. Many suitable thermistor types are available from different manufacturers.

In some temperature-sensing applications a suitable thermistor may not be readily available, or alternatively, conventional thermistors may not be suitable. In low-temperature applications, for example, a finite amount of power is dissipated in the thermister by the measuring circuitry, and this power dissipation may cause sufficient self-heating of the thermistor to upset its accuracy. In such cases a circuit of the type shown in *Figure 6.4*, in which a conventional silicon diode is used as a temperature-sensing element, may prove useful.

If a fixed current of a milliamp or so is passed through a conventional silicon diode a forward voltage of approximately 600 mV is developed across the junction of the device. The precise value of the forward voltage is subject to variation with temperature, and the junction in fact exhibits a negative-temperature-coefficient of approximately –2 mV/°C. All silicon diodes exhibit a similar temperature coefficient, and can thus be used as temperature-sensing devices. Since only 600 μW or so of power are dissipated in the diode at currents of 1 mA, self-heating effects are negligible, and the diode thus offers certain advantages over conventional thermistors in low-temperature sensing applications.

The operating principle of the *Figure 6.4* circuit is quite simple. Current is fed to zener diode ZD_1 via R_1, so a stabilised potential of 5.6 V is developed across ZD_1. This stabilised voltage is used to generate a constant current in silicon diode D_1 via R_4, and to generate an adjustable stabilised voltage in R_3 via R_2. Consequently, a temperature-dependent voltage is developed on the positive terminal of the op-amp, and a fixed reference potential is developed on the negative terminal. By adjusting the value of the reference potential via R_3 the *standing* voltage of the diode can be cancelled out to give a purely temperature-dependent differential voltage between the negative and positive terminals of the op-amp.

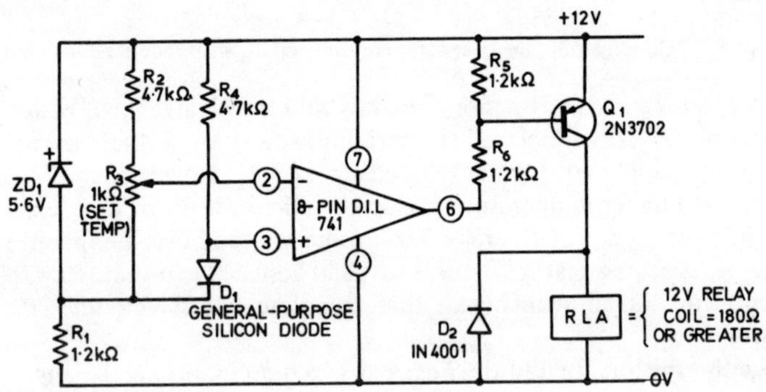

Figure 6.4. Over-temperature switch, using silicon diode temperature sensor.

Suppose then that R_3 is adjusted to give zero differential voltage at the required trip temperature. If the temperature now rises above the trip level the forward voltage of D_1 will fall, so the positive terminal of the op-amp will become negative relative to the negative terminal, and the op-amp will go into negative saturation. Under this condition, therefore, Q_1 and the relay are driven on. If, on the other hand, the temperature falls below the trip level, the D_1 voltage will rise, so the positive terminal of the op-amp will become positive relative to the negative terminal, and the op-amp will go into positive saturation. Under this condition Q_1 and the relay will be cut off.

Thus, the relay in the *Figure 6.4* circuit turns on only when the temperature exceeds the pre-set trip level. In practice, this circuit can be expected to give a thermal switching accuracy of better than 0.5 °C throughout the temperature range −50 °C to +120 °C. Temperature-sensing element D_1 can be any general-purpose silicon diode. The action of the circuit can be reversed, so that it acts as an under-temperature switch, by simply transposing the input terminal connections of the op-amp.

Figure 6.5 shows how a pair of silicon diodes can be used as temperature-sensing elements in a differential temperature switch, which turns on only when the temperature of D_2 is greater than that of D_1, and is not influenced by the absolute temperature of the two diodes. Circuit operation is as follows.

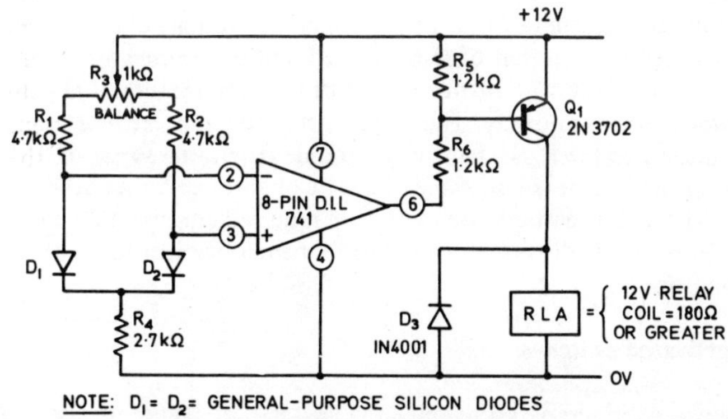

Figure 6.5. Differential temperature switch.

D_1 and D_2 are general-purpose silicon diodes, and are used as temperature-sensing elements. A standing current can be passed through D_1 from the positive supply line via R_3, R_1, and R_4, and a similar current can be passed through D_2 via R_3, R_2, and R_4. The relative values of these currents can be adjusted over a limited range via R_3, thus enabling the forward volt drops of the diodes to be equalised, so that they give zero differential output when they are both at the same temperature.

Suppose then that the diode voltages have been equalised in this way, so that zero voltage differential exists between them. If now the temperatures of both diodes are raised by 10 °C, the forward voltages of both diodes will fall by 20 mV, and zero voltage differential will still exist between them. The circuit is thus not influenced by identical changes in the ambient temperatures of D_1 and D_2. Suppose, on the other hand, that the temperature of D_2 falls 1 °C below that of D_1. In this case the D_2 voltage will rise 2 mV above that of D_1, so the positive terminal of the op-amp will go positive to the negative terminal, and the op-amp will go into positive saturation and hold Q_1 and the relay off. Finally, suppose that the temperature of D_2 rises 1 °C above that of D_1. In this case the D_2 voltage will fall 2 mV below that of D_1, so the positive terminal of the op-amp will go negative relative to the negative terminal, and the op-amp will go into negative saturation and drive Q_1 and the relay on. Thus, the relay turns on only when the temperature of D_2 is above that of D_1 (or when the temperature of D_1 is below that of D_2).

In the explanation above it has been assumed that R_3 is adjusted so that the D_1 and D_2 voltages are exactly equalised when the two diodes are at the same temperature, so that the relay goes on when the D_2 temperature rises a fraction of a degree above that of D_1. In practice, R_3 can readily be adjusted so that the standing bias voltage of D_2 is some millivolts greater than that of D_1 at normal ambient temperatures, in which case the relay will not turn on until the temperature of D_2 rises some way above that of D_1. The magnitude of this differential temperature trip level is in fact fully variable from zero to about 10 °C via R_3, so the circuit is quite versatile. The circuit can be set up by simply raising the temperature of D_2 the required amount above that of D_1, and then carefully adjusting R_3 so that the relay just turns on under this condition.

Light-activated switches

The precision temperature-activated switches of *Figures 6.1* to *6.3* can readily be made to function as light-activated switches by using light-dependent resistors (LDR) in place of the thermistors in their bridge networks. *Figure 6.6* shows the basic circuit of a light-activated switch, which turns on when the light intensity exceeds a pre-set level.

The LDR in this (and the *Figure 6.7* to *6.9* circuits) presents a low resistance under bright conditions, and a high resistance under dark conditions. Thus, under dark conditions the LDR resistance is high, so the voltage on the negative terminal of the op-amp is below that of the positive terminal, so that the op-amp is positively saturated and Q_1 and the relay are cut off. Under bright conditions, on the other hand, the LDR resistance is low, so the op-amp is negatively saturated, and Q_1 and the relay are driven on. The actual trip level of the circuit can be adjusted via R_1.

The action of the above circuit can be reversed, so that it functions as a dark-activated switch that turns on when the light intensity falls below a pre-set level, by merely transposing the R_1 and LDR positions, as shown in *Figure 6.7*. This circuit also shows how the design can be provided with electro-mechanical self-latching by wiring a pair of

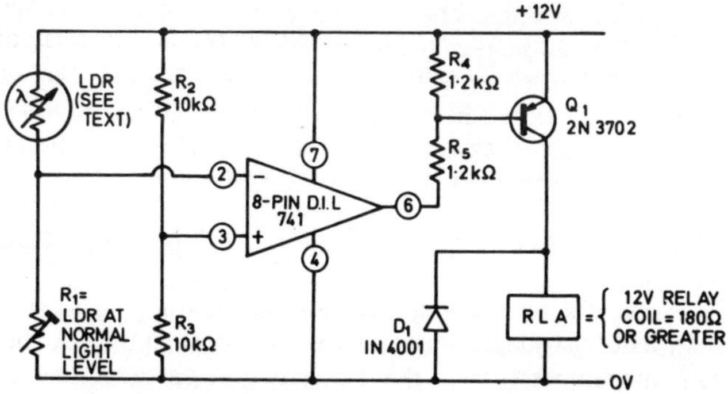

Figure 6.6. Precision light-activated switch.

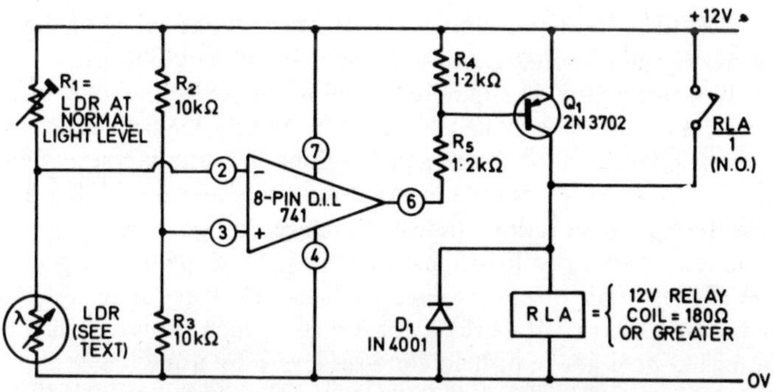

Figure 6.7. Self-latching dark-activated switch.

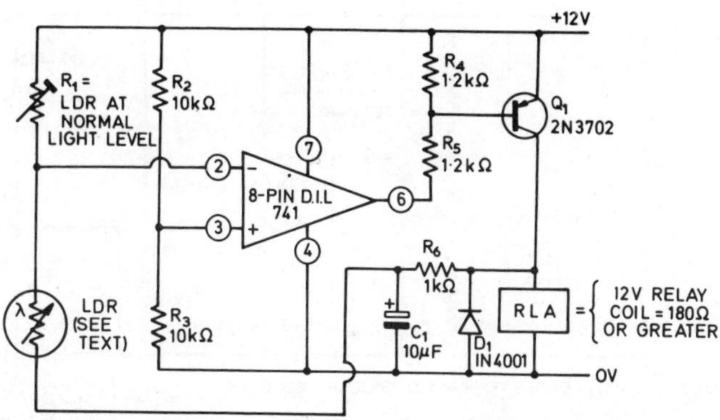

Figure 6.8. Dark-activated switch with electronic self-latching.

normally-open relay contacts between the emitter and collector of Q_1, so that the relay is automatically connected directly across the power supply lines once it has been initially activated by the sensing circuitry. Once the *Figure 6.7* circuit has self-latched into the ON mode, it can only be reset into the off state again by raising the light intensity above the trip level and momentarily breaking the supply connections to the relay.

The dark-operated switching circuit can be provided with a purely electronic self-latching facility, if required, by using the connections shown in *Figure 6.8*. Note in this case that the low end of the LDR is taken to the 0 V line via R_6 and the relay coil. Normally, when the light intensity is above the trip level, the relay is off, so the voltage on the negative terminal of the op-amp is dictated by the resistive values of the potential divider chain formed by R_1 – *LDR* – R_6 and the relay coil. Once the relay has turned on, however, 12 V is developed across its coil via Q_1, and this potential is imposed on the negative terminal of the op-amp via R_1 and the *LDR*, and thus locks the op-amp into the negatively saturated mode and holds the relay on. Once the relay has locked on, it can only be turned off again by raising the light intensity above the trip level, and momentarily breaking the supply connections to the circuit. Capacitor C_1 is used in this circuit to enhance stability and suppress the effects of sudden transient changes in light level.

A two-relay combined light/dark switch can be made by using an LDR in place of the thermistor in the *Figure 6.3* circuit. If required, the circuit can be modified so that it drives a single relay, which turns on when the light intensity goes above or below pre-set levels, by using the connections shown in *Figure 6.9*. In this case IC_1 acts as a dark-sensitive switch, and

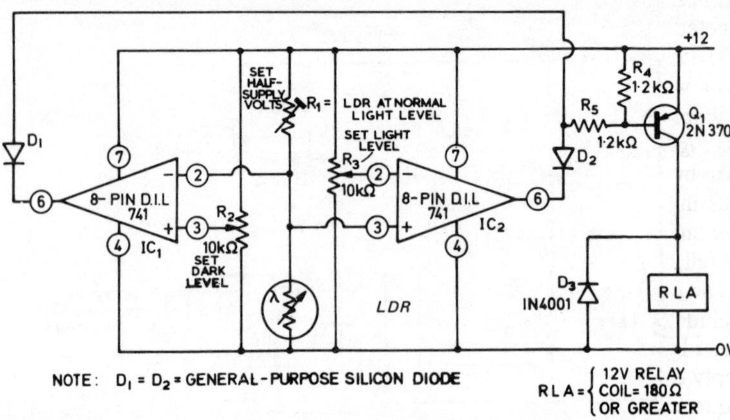

Figure 6.9. Combined light/dark switch with single input.

IC_2 as a light-sensitive switch, and the outputs of these two switches are fed to the relay via Q_1 and gating diodes D_1 and D_2 in such a way that the relay turns on whenever either op-amp goes into negative saturation. The procedure for setting up the *Figure 6.9* circuit is as follows.

First, set R_2 and R_3 to approximately mid travel, and then, with the LDR at its *normal* illumination level, adjust R_1 so that approximately half-supply volts are developed across the LDR. Now fully rotate the R_2 slider towards the positive supply line, and rotate the R_3 slider towards the 0 V line, and check that the relay is off. Next, reduce the LDR illumination intensity to the required *dark* trip level, and adjust R_2 so that the relay just turns on. Now increase the illumination level slightly, and check that the relay goes off. Finally, increase the illumination to the required *light* trip level, and adjust R_3 so that the relay again just turns on. All adjustments are then complete, and the unit is ready for use.

The LDRs used in the *Figures 6.6* to *6.9* circuits can be any cadmium sulphide photocells that present a resistance in the range 900 Ω to 900 kΩ at the required trip levels of light intensity.

A water or steam-activated switch

Figure 6.10 shows how the operational amplifier can be made to function as a sensitive water or steam-activated relay-driving switch. The op-amp is again wired as a voltage comparator, and has a fixed voltage

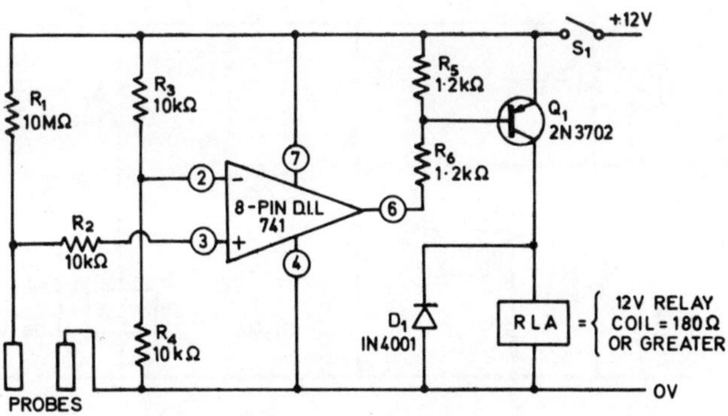

Figure 6.10. Water or steam-activated switch.

applied to its negative terminal via R_3 and R_4, and a variable voltage applied to the positive terminal via the potential divider formed by R_1 and the resistance of the two probes. If the probes are open circuit, negligible voltage divider action takes place, and almost the full supply line voltage is applied to the positive terminal of the op-amp: under this condition the relay is off. When, on the other hand, a resistance of less than 10 MΩ is connected across the probes, the voltage divider action is such that the voltage on the positive terminal of the op-amp is less than that of the negative terminal, and under this condition Q_1 and the relay turn on. Thus, the relay turns on whenever a resistance of less than 10 MΩ appears between the two probes.

Water and steam exhibit the characteristics of relatively low-value resistances. Consequently, when water or steam come into contact with both metal probes simultaneously, the relay turns on, and the circuit thus acts a water or steam-activated switch. The circuit has a variety of applications in the home and in industry. It can be used to sound an alarm when rain water falls on the probes, when flooding occurs in basements, when water rises to a pre-set level in tanks or baths, or when steam is ejected from a kettle spout as the liquid in the kettle starts to boil. If required, the sensitivity of the circuit can be reduced by increasing the value of R_1, since R_1 determines the maximum value of probe resistance that is needed to activate the circuit.

Voltage-activated switches

Operational amplifiers can readily be made to function as precision over-voltage or under-voltage switches in both d.c. and a.c. applications. *Figure 6.11*, for example, shows the connections for making a precision d.c. over-voltage switch for use with test voltages of 5 V or greater.

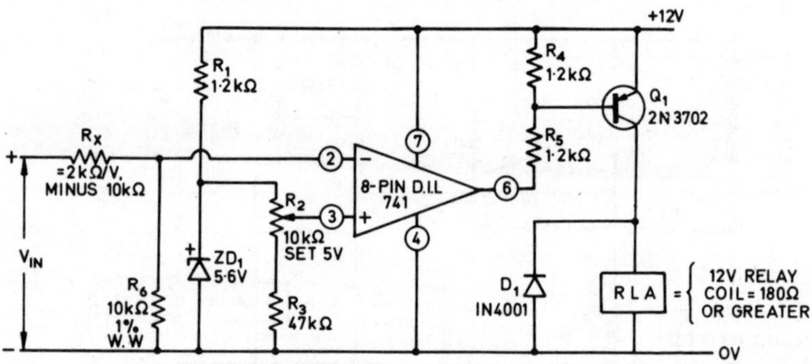

Figure 6.11. Precision d.c. over-volt switch, covering 5 V upwards.

The operation of the *Figure 6.11* circuit is quite simple. A zener-derived reference potential of 5 V is applied to the non-inverting pin of the op-amp via R_2, and the test voltage is applied between the inverting pin and ground via R_x. R_6 is a precision 10 kΩ resistor wired between the inverting pin and ground. The basic action of the circuit is such that the relay is off when the inverting pin voltage of the op-amp is less than the 5 V of the reference potential, and the relay is on when the inverting pin voltage is greater than 5 V.

Since R_x is wired in series between the actual test voltage and the 10 kΩ impedance of the inverting pin of the op-amp, R_x enables the circuit to be ranged so that it triggers at any required voltage in excess of the 5 V reference value. The value of R_x for any required trigger voltage value is in fact selected on a basis of 2 kΩ/V – 10 kΩ. Thus, for 50 V triggering, R_x = 50 x 2 kΩ – 10 kΩ = 90 kΩ. For 5 V triggering, R_x should be given a value of zero.

The *Figure 6.11* circuit is very sensitive, and exhibits negligible backlash; triggering accuracies of 0.5 % can easily be achieved. For maximum accuracy, the zener reference supply of the circuit should itself be stabilised, or the zener should be fed from a constant-current source.

The *Figure 6.11* circuit can be made to function as a precision under-voltage switch, which turns on whenever the input voltage falls below a pre-set level, by simply transposing the inverting and non-inverting pin connections of the op-amp, as shown in *Figure 6.12*. This circuit also

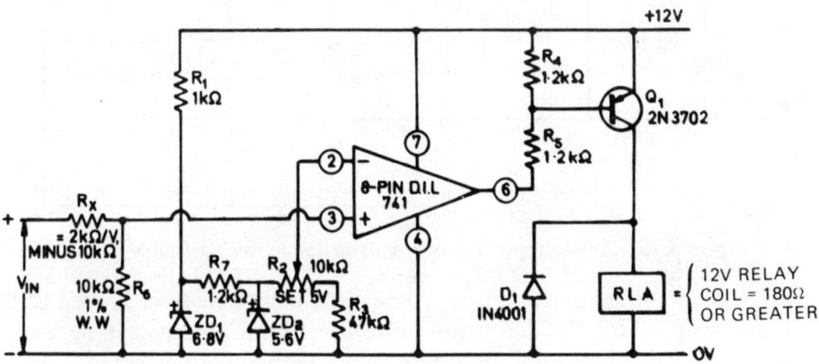

Figure 6.12. **Precision d.c. under-volt switch, covering 5 V upwards.**

shows how the zener reference supply can be stabilised for high-precision operation. Note in both of these circuits that, once 5 V has been accurately set via R_2, the final triggering accuracy of each design is determined solely by the accuracies of R_x and R_6. In high-precision applications, therefore, both R_x and R_6 should be precision wire-wound types.

Figure 6.13 shows how the *Figure 6.11* circuit can be modified for use as an over-voltage switch covering the input voltage range 10 mV to 5 V. In this case the input voltage is connected directly to the inverting terminal of the op-amp, and a variable reference potential is applied to the non-inverting terminal. This reference potential is adjusted to give the same value as that of the required trigger voltage. Note that this circuit makes use of two sets of supply lines, to ensure proper biasing of the op-amp.

Figure 6.14 shows how the above circuit can be modified for operation from a single set of supply lines. In this case the zener diode is given a value of 6.8 V, and R_1 is connected in series with ZD_1 via the negative supply line. Note that the $R_1 - ZD_1$ junction forms the common input signal connection of the op-amp. Consequently, the positive supply terminal of the op-amp is biased 6.8 V positive to the common line, and the negative supply terminal is biased 5.2 V to 11.2 V negative to the common line. The op-amp is thus correctly biased even though only a single set of supply lines is used.

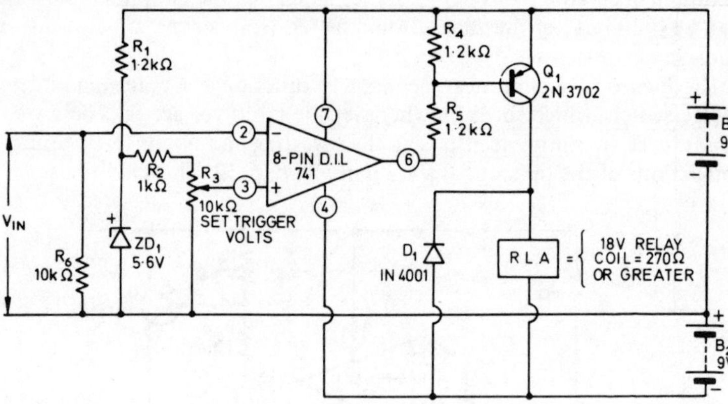

Figure 6.13. Dual supply d.c. over-volt switch, covering 10 mV to 5 V.

The four voltage-activated switching circuits that we have looked at so far are designed for d.c. activation only. These basic circuits can be modified for a.c. activation by interposing suitable rectifier/smoothing networks between their input terminals and the actual a.c. input signals,

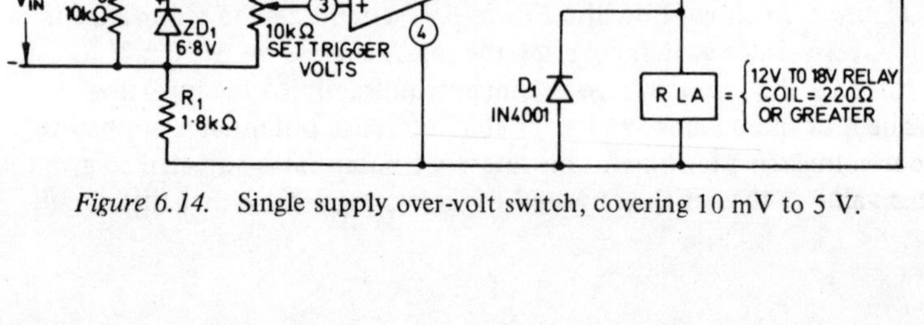

Figure 6.14. Single supply over-volt switch, covering 10 mV to 5 V.

so that the a.c. signals are converted to d.c. before being applied to the basic circuits.

Figure 6.15a shows the practical circuit of a precision a.c. over-voltage switch that is designed to work with sine wave signals in excess of 2.5 V r.m.s. Here, the a.c. signal is converted to d.c. via voltage doubling and smoothing network $R_x - C_1 - D_1 - D_2 - C_2 - R_1$, and the resulting d.c. voltage is applied to the inverting input of the op-amp via R_2; a zener-derived 5.6 V reference potential is applied to the non-inverting terminal. The circuit action is such that the relay turns on only when the d.c. voltage on the inverting terminal exceeds 5.6 V.

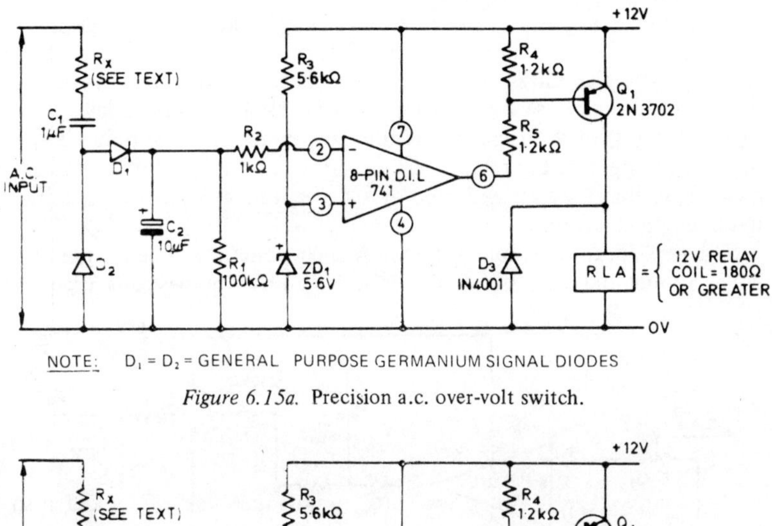

NOTE: $D_1 = D_2$ = GENERAL PURPOSE GERMANIUM SIGNAL DIODES

Figure 6.15a. Precision a.c. over-volt switch.

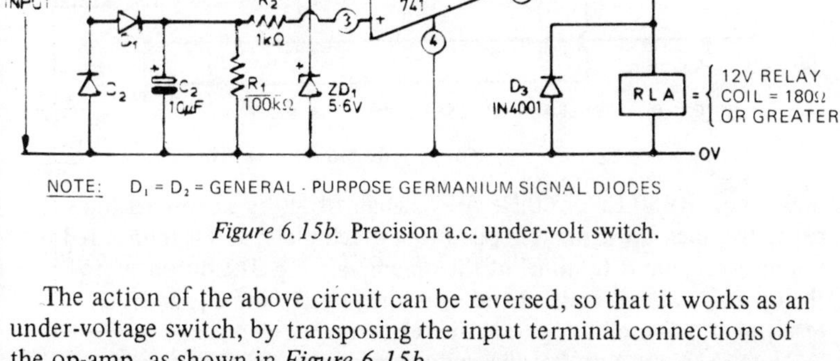

NOTE: $D_1 = D_2$ = GENERAL-PURPOSE GERMANIUM SIGNAL DIODES

Figure 6.15b. Precision a.c. under-volt switch.

The action of the above circuit can be reversed, so that it works as an under-voltage switch, by transposing the input terminal connections of the op-amp, as shown in *Figure 6.15b*.

It should be noted that both of the *Figure 6.15* circuits exhibit a basic input impedance, with R_x reduced to zero ohms, of approximately 15 kΩ, and that under this condition a sine wave input of roughly 2.5 V r.m.s. is needed to operate the relay. Consequently, when R_x is given a finite value it acts as a potential divider with this 15 kΩ input impedance, and enables the circuit to be ranged to trigger at any required a.c. voltage level in excess of 2.5 V. The R_x value should be chosen on the basis of approximately 6 kΩ/V – 15 kΩ.

Miscellaneous switching projects

Operational amplifiers can be made to function as very sensitive a.c. over-volt switches in a variety of ways. One method is to use the op-amp as an open-loop amplifier of a.c. input signals, and to then rectify the op-amp output and use the resulting d.c. to drive a transistor-relay switch. Circuits of this type may require a.c. input signals of only a few millivolts to operate the relay. Such circuits can be adapted in a number of ways, so that they are activated by specific frequencies, or by specific frequency bands, etc.

Figure 6.16, for example, shows how such a circuit can be adapted for use as a touch-operated switch that is activated by a bodies capacitive

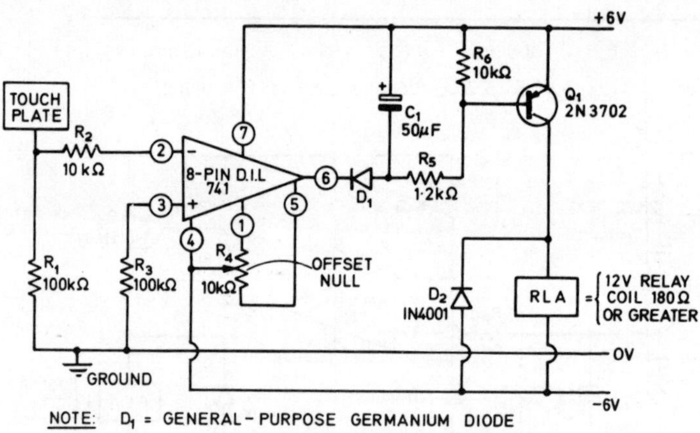

Figure 6.16. Touch-activated relay switch.

pick-up of the 50 Hz or 60 Hz stray signals from the a.c. power lines. The capacitive pick-up signal is applied to a touch plate, and is thence fed to the inverting input terminal of the op-amp via R_2. The output signal of the op-amp is then rectified and smoothed by D_1 and C_1, and the resulting d.c. is used to operate the relay via Q_1.

Note in this circuit that the rectifier and relay-driving network are d.c.

coupled to the output of the op-amp, and that the op-amp is provided with an offset-null control. In practice, the offset-null control is adjusted so that the op-amp output is just short of positive saturation under quiescent conditions. Under these circumstances a small d.c. bias is applied to the base of Q_1 via D_1 and R_5, but is not sufficient to drive Q_1 or the relay on. Consequently, when a.c. signals are applied to the op-amp, the negative-going parts of the output signals apply sufficient extra biasing to drive Q_1 and the relay on, and the circuit exhibits very high sensitivity. When the offest-null control is correctly adjusted, the relay can be activated by input signals as low as 1 mV r.m.s.

The touch plate in the *Figure 6.16* circuit can be a simple metal disc a few centimetres in diameter. Note that the common rail of the circuit must be grounded for correct operation, and that the circuit will only work if it is within reasonable proximity of the a.c. power lines. To set up the circuit, first adjust R_4 (the offset-null control) so that the relay goes on, and then turn R_4 back just past the point at which the relay goes off again. Now touch the touch plate, and check that the relay turns on. All adjustments are then complete, and the circuit is ready for use.

Figure 6.17 shows how the above circuit can be modified so that it acts as a voice-frequency relay switch. The two circuits are similar, except that in the case of the *Figure 6.17* circuit the op-amp is wired as a

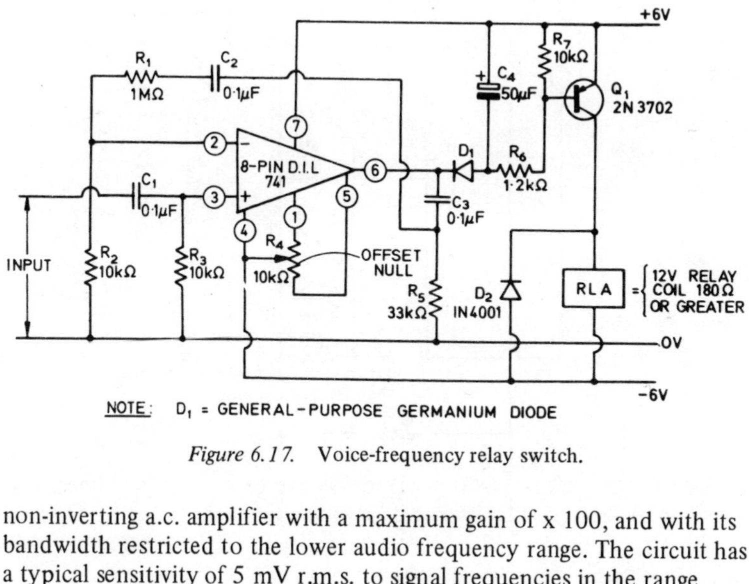

Figure 6.17. Voice-frequency relay switch.

non-inverting a.c. amplifier with a maximum gain of x 100, and with its bandwidth restricted to the lower audio frequency range. The circuit has a typical sensitivity of 5 mV r.m.s. to signal frequencies in the range 50 Hz to 2 kHz, and a sensitivity of 10 mV over the range 50 Hz to 4.5 kHz.

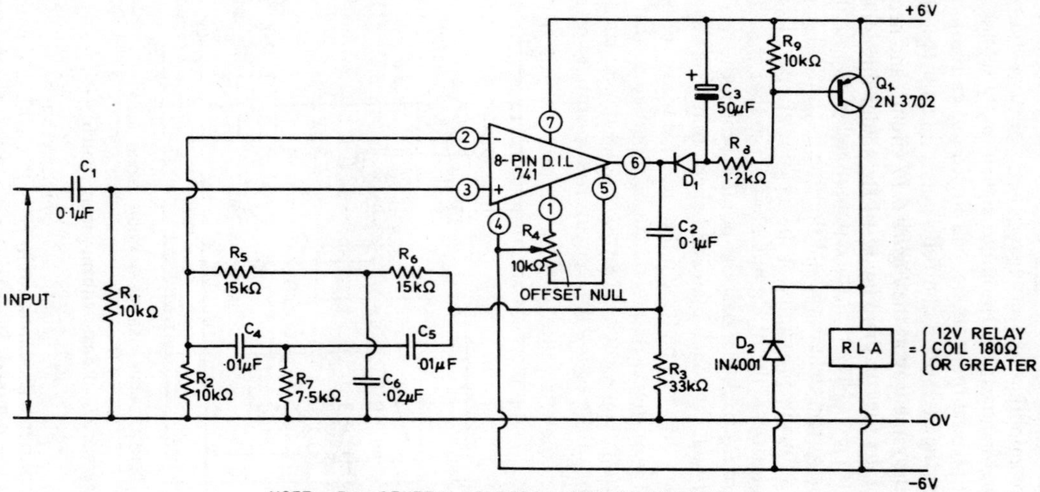

NOTE: D_1 = GENERAL-PURPOSE GERMANIUM DIODE

Figure 6.18. 1 kHz tone-activated relay switch.

The *Figure 6.17* circuit is set up by first adjusting R_4 so that the relay goes on, and then turning R_4 back so that the relay just goes off again. The relay should then operate if a 1 kHz signal of a 5 mV or so is applied to the circuits input. The circuit can be used as a sound-operated switch, if required, by simply feeding the output of a pick-up microphone to the unit via a simple pre-amplifier, to give the required 5 mV drive voltage.

Finally, *Figure 6.18* shows how the above circuit can be modified for use as a 1 kHz tone-operated relay switch by incorporating a twin-T filter network in the negative feedback path of the non-inverting amplifier. The circuit is set up in the same way as the *Figure 6.17* circuit, and needs an input of 5 mV r.m.s. to operate the relay at 1 kHz: under this condition the circuit bandwidth is roughly ± 2.5 % of the centre frequency. To maintain this narrow bandwidth, the input signal must be limited to less than 10 mV r.m.s., since the bandwidth is proportional to the amplitude of the input signal of the circuit.

Time-activated switches

Finally, to conclude this last chapter of this volume, *Figures 6.19* and *6.20* show how op-amps can be used as time-activated switches with relay outputs. The *Figure 6.19* circuit acts as a 100 s delayed-turn-on relay driver, in which the relay does not turn on until 100 s after S_1 is closed. The *Figure 6.20* circuit acts as an auto-turn-off relay driver, in which the relay turns on as soon as S_1 is operated, but turns off again automatically after a delay of 100 s.

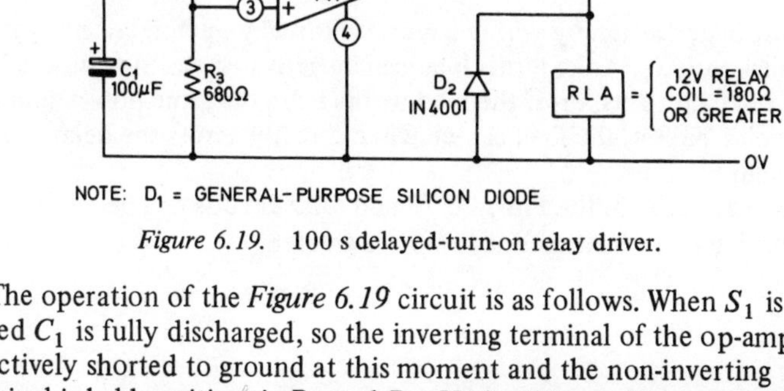

NOTE: D_1 = GENERAL-PURPOSE SILICON DIODE

Figure 6.19. 100 s delayed-turn-on relay driver.

The operation of the *Figure 6.19* circuit is as follows. When S_1 is first closed C_1 is fully discharged, so the inverting terminal of the op-amp is effectively shorted to ground at this moment and the non-inverting terminal is held positive via R_2 and R_3. Under this condition the op-amp output is driven to saturation, so Q_1 and the relay are both cut off. C_1 starts to charge exponentially via R_1 as soon as S_1 is closed, and after a pre-set delay of about 100 s the inverting terminal voltage rises above that of the non-inverting terminal, at which point the op-amp comes out of positive saturation and Q_1 and the relay turn on. The operating sequence is then complete.

The *Figure 6.19* circuit gives time delays of approximately 1 s per μF of C_1 value, and thus gives a delay of 100 s with the component values shown. If required, the delay can be made variable by replacing R_1 with a 1 MΩ variable resistor. D_1 is used in the circuit to rapidly discharge C_1 via the low value resistance of R_2 and R_3 when S_1 is opened, thus giving a rapid reset action.

The action of the *Figure 6.19* circuit can be reversed, so that it gives automatic-turn-off operation, by simply transposing the inverting and non-inverting terminal connections of the op-amp, as shown in *Figure 6.20*. This diagram also shows how the circuit can be adapted for

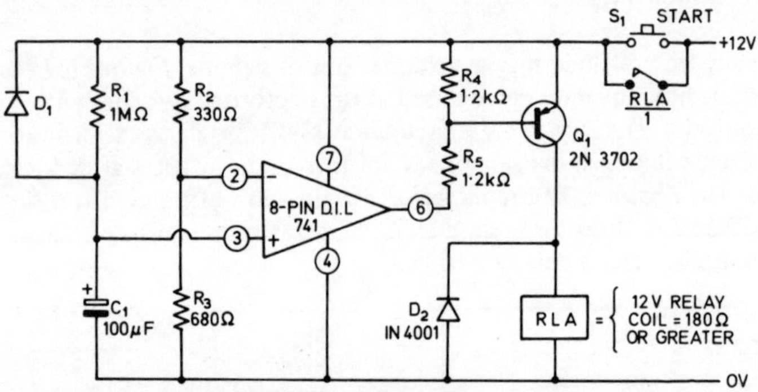

Figure 6.20. **Pushbutton activated auto-turn-off relay driver.**

pushbutton activation by wiring a pair of normally-open relay contacts in parallel with S_1. An external load can be activated via an additional pair of relay contacts, or, if the load works from the same power supply as the delay circuit, the load can be wired directly across the delay circuits supply lines.

The relays used in the *Figure 6.19* and *6.20* circuits can be any 12 V types having coil resistances of 180 Ω or greater.

APPENDIX

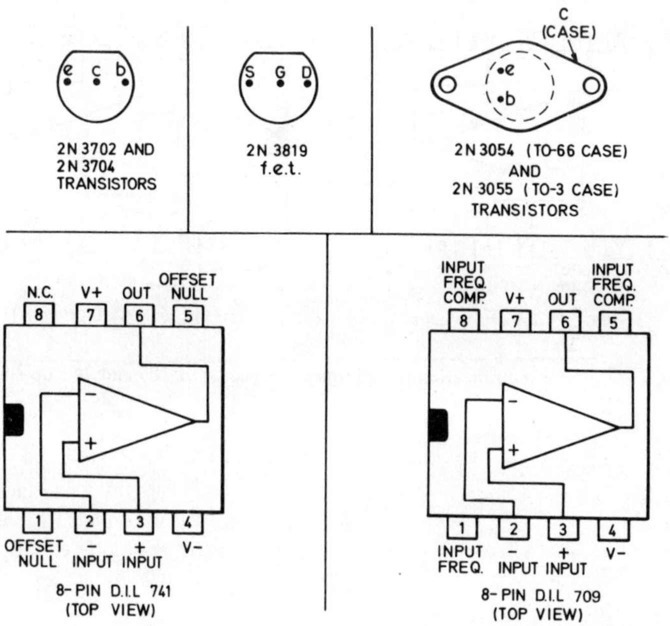

Figure 7.1. Outlines and connections of semiconductor devices used in the volume.

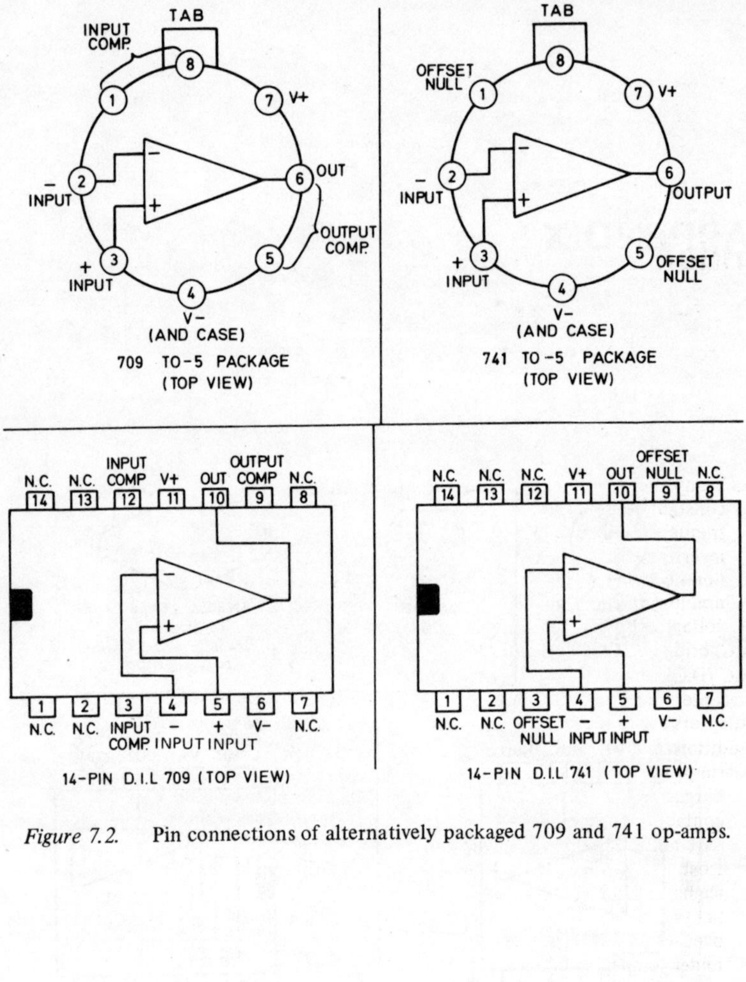

Figure 7.2. Pin connections of alternatively packaged 709 and 741 op-amps.

Index